Rice Quality Handbook

Randall G. Mutters
Farm Advisor
University of California Cooperative Extension
Agriculture and Natural Resources, Central Valley Region

James F. Thompson
Agricultural Engineering Specialist
University of California Cooperative Extension
University of California, Davis

University of California
Agriculture and Natural Resources
Publication 3514

To order or obtain ANR publications and other products, visit the ANR Communication Services online catalog at *http://anrcatalog.ucdavis.edu* or phone 1-800-994-8849. You can also place orders by mail or FAX, or request a printed catalog of our products from

University of California
Agriculture and Natural Resources
Communication Services
6701 San Pablo Avenue, 2nd Floor
Oakland, California 94608-1239
Telephone 1-800-994-8849
510-642-2431
FAX 510-643-5470
E-mail: danrcs@ucdavis.edu

Publication 3514
ISBN-13: 978-1-60107-622-9
Library of Congress Control Number: 2009902340

Photo credits are given in the captions. Photo on page i by Michael Poe.
Cover photographs by Michael Poe; design by Celeste A. Rusconi.

To simplify information, trade names of products have been used. No endorsement of named or illustrated products is intended, nor is criticism implied of similar products that are not mentioned or illustrated.

This publication has been anonymously peer reviewed for technical accuracy by University of California scientists and other qualified professionals. This review process was managed by the ANR Associate Editor for Agronomy and Range Sciences.

1m-pr-10/09-SB/CR

Contents

Preface

This book provides practical information on how to produce high-quality rice in California. It discusses the quality implications of crop planning, cultural practices, harvest procedures, drying methods, and storage systems. Growers and operators of dryers, warehouses, and processing operations will benefit from this information.

This book describes rice growing and handling in the Sacramento Valley in Northern California. This region produces mainly medium and short grain rice in a flooded culture. Rice varieties are developed by local breeders and are specifically adapted to the microclimates in the valley to produce a quality suitable for the industry's traditional customers. The valley has a Mediterranean climate characterized by very little rainfall during the growing season. Humidity is low and daily air temperature minimums and maximums differ by about 40°F. Rice is harvested mostly in September and October. Periods of strong, dry north winds lasting from 2 to 5 days are common during harvest, especially in October. Winters are cool, but temperatures rarely drop below about 30°F. Rain and fog are common from mid-November through February.

Much of the information in this book is based on information developed by University of California Cooperative Extension; University of California, Davis, faculty; and USDA researchers at the Western Regional Research Laboratory in Albany, California. We would like to thank Jack Williams for writing an early version of the chapter on harvesting. We want to especially thank George E. Miller Jr., who spent many years serving the rice industry and who inspired our interest in improving rice quality.

We appreciate the California Rice Research Board's generous funding of most of our research on rice quality. The California Rice Commission has funded our annual Rice Quality Workshop since its inception in 1999 and provided the funds to print this book.

California Rice Production Region

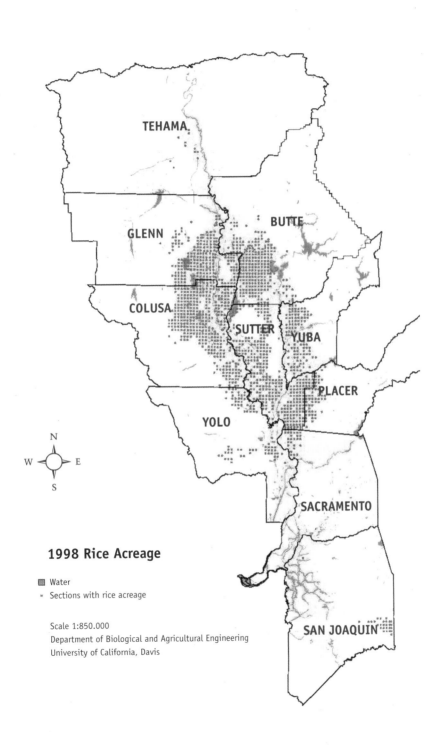

1998 Rice Acreage

▨ Water
▪ Sections with rice acreage

Scale 1:850.000
Department of Biological and Agricultural Engineering
University of California, Davis

CHAPTER 1

Rice Quality in the Global Market

During the past 25 years there has been a pronounced shift in demand from low-quality to high-quality rice as a result of rising incomes and market liberalization. In the mid-1970s, low-quality rice comprised 38 percent of the world trade, but that percentage declined to just 23 percent in 2006. Generally, improvement in economic conditions leads to a shift in the demand from low-quality to high-quality rice. Consumers do not purchase more rice but instead purchase better-quality rice. These trends are evident in Japan as well as in fast-growing countries such as China, South Korea, Thailand, and Malaysia. Consumers are increasingly reluctant to accept low-quality rice as a substitute for rice of high quality.

About 5 percent of the world's production of paddy rice is traded on the world market; 95 percent of rice is domestically consumed by the country of origin. Seven countries account for 90 percent of rice exports worldwide (fig. 1.1). The U.S. exports represented 10 percent of the total traded rice in 2007, the fourth-largest following Thailand, India, and Vietnam. Sales of long grain rice to Latin America are a major portion of the U.S. export market. The small amount of rice traded relative to the total amount produced creates the potential for highly variable world prices because production changes in a few countries may cause substantial shifts in exportable supplies.

In 2007, world trade of rice was dominated by indica types, representing 74 percent of the market (fig. 1.2). Japonica and aromatic types contributed 16 percent and 9 percent, respectively, and the remaining 1 percent was glutinous rice. Virtually all of the rice grown in California is japonica, predominately medium grain varieties. California production accounts for about 65 percent of the total U.S. medium grain production. In 2002, about 50 percent of California production was used domestically either by direct consumption or processing; use in beer accounts for about 16 percent of the domestic market. The remainder, about 50 percent, was exported. In 2006, northern Asia was the largest market for California rice, principally Japan, Korea, and Taiwan. Japan is California's largest export customer due in part to the General Agreement on Trade and Tariffs (GATT). Rice grown in the United States, mainly in California, accounts for nearly 50 percent of Japan's World Trade Organization commitment to import 8 percent of its annual consumption of rice. On average from 1995 to 2005, California delivered 48 percent of all the rice imported in Japan. Japan accounts for roughly half of all California rice export sales, or the equivalent of between 20 percent and 25 percent of California's annual rice production. In Korea, California supplies about 85 percent of the rice that is imported from the United States. California's major competitors in the world japonica market are northern China and Egypt. The cost of rice production can be less in these countries than in California. Consequently, California's competitive edge and ultimate market retention depends to a considerable degree on the consistent production and delivery of high-quality rice.

Figure 1.1. Major rice-exporting countries and their corresponding share of the 31 million metric tons of rice traded on the world market in the 2006/2007 production year (%). *Source:* FAS 2007.

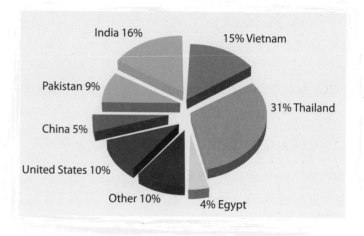

Figure 1.2. The major categories of rice traded on the world market and their corresponding market shares (%). *Source:* FAS 2007.

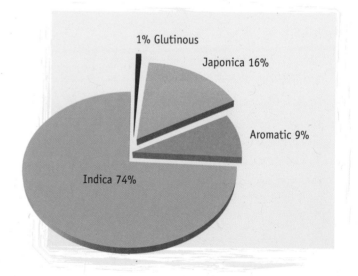

Definitions of Quality

The diverse domestic and export markets require that quality be evaluated according to specific end uses. Whether rice is acceptable for a particular use is determined by a variety of subjective and objective factors. How these features are ranked in importance depends largely on the end user. Rice, unlike most other cereal grains, is consumed primarily as a whole grain. Therefore, physical properties of the milled rice, such as size, shape, uniformity, and general appearance, are of utmost importance.

Rice quality is partially determined by the genetic makeup of a variety. Selection for improved milling, cooking, eating, and processing qualities is an essential component of rice breeding programs. The goal of these programs is to develop rice cultivars that meet industry standards for the end use characteristics preferred by consumers in specific domestic and export markets.

Environmental conditions and cultural practices during growth also produce marked differences in rice quality (table 1.1). These factors can have a greater effect on quality than can inherited traits. For example, the rehydration of low-moisture kernels by heavy dew or rain following a period of north wind can cause a substantial reduction in milling quality in some varieties. Furthermore, quality is lost during improper handling and storage or by the presence of foreign material. Physically moving the rice can damage the kernels and decrease milling quality. Prolonged periods of storage under conditions of high moisture content can result in objectionable flavors or odors. Undried rice exposed to prolonged periods of improper aeration develop off odors, whether sitting in trucks awaiting transport or in temporary storage at the dryer.

Measures of quality vary somewhat between paddy and milled rice. These characteristics are discussed in general terms below. However, specific standards for each characteristic may vary by country, organization, or end use. The U.S. Department of Agriculture (USDA) has established standards for rice quality assessments and

Table 1.1. Genetic and cultural components of rice quality in the United States

Component	Genetic	Cultural
milling quality	X	X
hull and bran color	X	
grain shape	X	X
cooking characteristics	X	X
moisture content		X
test weight	X	X
color	X	X
dockage		X
damaged grains		X
odors		X
presence of red rice		X

Key Concepts in Rice Quality

- Cultural practices and growing conditions control initial quality at harvest.
- Improper drying and handling procedures after harvest result in off odors and poor milling quality.
- The quality of rice changes as the crop ages in storage.
- Changes in quality are faster at higher grain temperatures and higher moisture contents.
- The intended use or market determines the desirable quality characteristics.

handling practices, and it oversees a network of public and private laboratories that provide user-fee-funded inspection and grading services. This program, referred to as the Federal Grain Inspection Service (FGIS), provides producers, marketers, and end users accurate descriptions of the quality of rice bought and sold in the United States. The FGIS has published grade standards for rough rice (table 1.2), brown rice (table 1.3), and milled rice (table 1.4). The USDA sample grade, and therefore the market value, of these three forms of rice depends principally on moisture content, physical characteristics, contaminates, and the presence of insects and microbially damaged kernels. An objectionable odor reduces grade. The proper method for preparation and handling of a rice sample submitted for inspection and grade is discussed in chapter 7.

Table 1.2. USDA grades and grade requirements for rough rice (§868.210; see also §868.212)

| | Maximum limits of - - - | | | | | | | |
| | Seeds and heat-damaged kernels | | | | Chalky kernels[1,2] | | | |
Grade	Total (singly or combined) (number in 500 grams)	Heat-damaged kernels and objectionable seeds (singly or combined) (number in 500 grams)	Heat-damaged kernels (number in 500 grams)	Red rice and damaged kernels (singly or combined) (percent)	In long grain rice (percent)	In medium or short grain rice (percent)	Other types[3] (percent)	Color requirements[1] (minimum)
U.S. No. 1	4	3	1	0.5	1.0	2.0	1.0	Shall be white or creamy
U.S. No. 2	7	5	2	1.5	2.0	4.0	2.0	May be slightly gray
U.S. No. 3	10	8	5	2.5	4.0	6.0	3.0	May be light gray
U.S. No. 4	27	22	15	4.0	6.0	8.0	5.0	May be gray or slightly rosy
U.S. No. 5	37	32	25	6.0	10.0	10.0	10.0	May be dark gray or rosy
U.S. No. 6	75	75	75	15.0[4]	15.0	15.0	10.0	May be dark gray or rosy

Revised December 1, 2002, Updated July 2005
U.S. Sample grade shall be rough rice which: (a) does not meet the requirements for any of the grades from U.S. No. 1 to U.S. No. 6, inclusive; (b) contains more than 14.0 percent moisture; (c) is musty, or sour, or heating; (d) has any commercially objectionable foreign odor; or (e) is otherwise of distinctly low quality.
[1] For the special grade Parboiled rough rice, see §868.212(b).
[2] For the special grade Glutinous rough rice, see §868.212(d).
[3] These limits do not apply to the class Mixed Rough Rice.
[4] Rice in grade U.S. No. 6 shall contain not more than 6.0 percent of damaged kernels.
[56 FR 55978, Oct. 31, 1991]

Source: Reprinted from FGIS 2005, p. 11.

Table 1.3. USDA grades and grade requirements for brown rice (short, medium, and long grain) (§868.261; see also §868.263)

	Maximum limits of ---									
	Paddy kernels		Seeds and heat-damaged kernels			Red rice and damaged kernels (singly or combined) (percent)	Chalky kernels[1,2] (percent)	Broken kernels removed by a 6 plate or a 6½ sieve[3] (percent)	Other Types[4] (percent)	Well-milled kernels (percent)
Grade	(percent)	(number in 500 grams)	Total (singly or combined) (number in 500 grams)	Heat-damaged kernels (number in 500 grams)	Objectionable seeds (number in 500 grams)					
U.S. No. 1	—	20	10	1	2	1.0	2.0	1.0	1.0	1.0
U.S. No. 2	2.0	—	40	2	10	2.0	4.0	2.0	2.0	3.0
U.S. No. 3	2.0	—	70	4	20	4.0	6.0	3.0	5.0	10.0
U.S. No. 4	2.0	—	100	8	35	8.0	8.0	4.0	10.0	10.0
U.S. No. 5	2.0	—	150	15	50	15.0	15.0	6.0	10.0	10.0

Revised December 1, 2002, Updated July 2005

U.S. Sample grade shall be brown rice for processing which (a) does not meet the requirements for any of the grades from U.S. No. 1 to U.S. No. 5, inclusive; (b) contains more than 14.5 percent of moisture; (c) is musty, or sour, or heating; (d) has any commercially objectionable foreign odor; (e) contains more than 0.2 percent of related material or more than 0.1 percent of unrelated material; (f) contains 2 or more live weevils or other live insects; or (g) is otherwise of distinctly low quality.

[1] For the special grade Parboiled brown rice for processing, see §868.263(a).
[2] For the special grade Glutinous brown rice for processing, see §868.263(c).
[3] Plates should be used for southern production rice and sieves should be used for western production rice, but any device or method which gives equivalent results may be used.
[4] These limits do not apply to the class Mixed Brown Rice for Processing.
[56 FR 55979, Oct. 31, 1991]
Source: Reprinted from FGIS 2005, p. 18.

Table 1.4. USDA grades and grade requirements for milled rice (short, medium, and long grain) (§868.310; see also §868.315)

	Maximum limits of ---											Color requirement[1] (minimum)	Milling requirement[5] (minimum)
	Seeds, heat-damaged, and paddy kernels (singly or combined)		Red rice and damaged kernels (singly or combined) (percent)	Chalky kernels[1,2]		Broken kernels				Other types[4]			
Grade	Total (number in 500 grams)	Heat-damaged kernels and objectionable seeeds (number in 500 grams)		In long grain rice (percent)	In medium or short grain rice (percent)	Total (percent)	Removed by a 5 plate[3] (percent)	Removed by a 6 plate[3] (percent)	Through a 6 sieve[3] (percent)	Whole kernels (percent)	Whole and broken kernels (percent)		
U.S. No. 1	2	1	0.5	1.0	2.0	4.0	0.04	0.1	0.1	—	1.0	Shall be white or creamy	Well milled
U.S. No. 2	4	2	1.5	2.0	4.0	7.0	0.06	0.2	0.2	—	2.0	May be slightly gray	Well milled
U.S. No. 3	7	5	2.5	4.0	6.0	15.0	0.1	0.8	0.5	—	3.0	May be light gray	Reasonably well milled
U.S. No. 4	20	15	4.0	6.0	8.0	25.0	0.4	1.0	0.7	—	5.0	May be gray or slightly rosy	Reasonably well milled
U.S. No. 5	30	25	6.0[5]	10.0	10.0	35.0	0.7	3.0	1.0	10.0	—	May be dark gray or rosy	Reasonably well milled
U.S. No. 6	75	75	15.0[6]	15.0	15.0	50.0	1.0	4.0	2.0	10.0	—	May be dark gray or rosy	Reasonably well milled

Revised December 1, 2002, Updated July 2005

U.S. Sample grade shall be milled rice of any of these classes which: (a) does not meet the requirements for any of the grades from U.S. No.1 to U.S. No.6, inclusive; (b) contains more than 15.0 percent of moisture; (c) is musty or sour, or heating; (d) has any commercially objectionable foreign odor; (e) contains more than 0.1 percent of foreign material; (f) contains two or more live or dead weevils or other insects, insect webbing, or insect refuse; or (g) is otherwise of distinctly low quality.

[1] For the special grade Parboiled milled rice, see §868.315(c).
[2] For the special grade Glutinous milled rice, see §868.315(e).
[3] Plates should be used for southern production rice; and sieves should be used for western production rice, but any device or method which gives equivalent results may be used.
[4] These limits do not apply to the class Mixed Milled Rice.
[5] For the special grade Undermilled milled rice, see §868.315(d).
[6] Grade U.S. No. 6 shall contain not more than 6.0 percent of damaged kernels.
[56 FR 55979, Oct. 31, 1991; 67 FR 61250, Sept. 30, 2002, as amended at 70 FR 37255, June 29, 2005]
Source: Reprinted from FGIS 2005, p. 27.

Internationally, the Codex Alimentarius (Latin for "food code") is the global food quality reference for international trade, national food agencies, and processors. Codex Alimentarius was created in 1963 by the Food and Agricultural Organization of the United Nations (FAO) and the World Health Organization (WHO) to develop food standards with the intent of protecting the health of consumers, ensuring fair practices in the food trade, and promoting the coordination of international food standards. Currently, the United States and 172 other countries plus the European Community recognize these standards of quality. Grain inspection according to Codex Alimentarius standards does not assign a grade to the rice, as does the USDA system. It establishes a minimum standard of quality suitable for trade among the participating countries. In addition to the quality characteristics in the USDA system, levels of heavy metals, pesticide residues, and packaging are also considered. Generally, ISO (International Organization for Standardization) analytical procedures are used for inspection and certification under the Codex Alimentarius system.

USDA Rice Inspection Handbook

A complete description of the policies and procedures for sampling, inspecting, and certificating rice in accordance with USDA regulations are contained in the Rice Inspection Handbook. It can be downloaded at http://www.gipsa.usda.gov/GIPSA/webapp?area=home&subject=lr&topic=hb-ri.

Paddy rice with intact husk (left), brown rice with husk removed (center), and milled rice with the bran and germ removed (right). *Photo: Michael Po*

General Categories of Rice

Rice is primarily classified according to its kernel form and grain shape into three broad categories: long, medium, and short grains. There are differences in cooking qualities within the grain shape categories, although the long grain varieties are typically firm and not sticky when cooked, while medium and short grain varieties are soft and sticky.

Kernel Forms

Rough, or paddy, rice

Rough, or paddy, rice is the harvested rice as it comes out of the combine with an intact hull, or husk. The rice in this form is dried and stored prior to milling. The rice hull, removed before milling, is frequently sold as animal feed or animal bedding or is burned as an energy source.

Brown rice

Brown rice is produced when the husk is removed from the paddy rice (for an illustration of the parts of a rice kernel, see fig. 10.3, p. 126). The bran layer and germ impart the color to the unmilled rice. The color of brown rice may vary from light tan to dark purple (sometimes referred to as black rice), depending on the variety. Milling removes the bran and germ that contain the greater amounts of lipids, vitamins, minerals, and fiber than the endosperm (the white center of the kernel). Shelling the rice to remove the husk releases enzymes from the bran layer that break down lipids. This breakdown makes brown rice more susceptible to becoming rancid than paddy rice. Rancidity imparts undesirable odors and flavors to rice. It is advisable to store brown rice under cool conditions to prolong its shelf life.

Comparison of Grain Shapes

Milled short (left), medium (center), and long grain rice (right).
Photo: Jack Kelly Clark.

White, polished, or milled rice

White, polished, or milled rice has had its bran removed, leaving only the endosperm of the kernel. Table rice, and that used in industrial processes such as cereal production, is overwhelmingly milled rice. In the United States, most white rice packaged for direct consumption is sometimes fortified with iron, niacin, thiamin, and folic acid.

Grain Shape

Long grain

Long grain rice is approximately three times longer than it is wide (table 1.5). Typical U.S. long grain rice has an intermediate gelatinization temperature and an apparent amylose content of 19 to 23 percent. Amylose is a polymer of several thousand glucose units. It is one of the two components of starch, the other being amylopectin. Amylose starch is the dominant storage form in plants, making up about 30 percent of the stored starch in plants. California long grain rice has a slightly higher apparent amylose content and lower gelatinization temperature than does long grain rice grown in the southern United States. In general, most consumers in the Western Hemisphere, southern China, and Europe prefer long grain rice.

Medium grain

Medium grain rice is approximately 2.5 times longer than it is wide (table 1.5). California is the largest U.S. producer of medium grain rice, and about 90 percent of the California production is medium grain. Medium grain rice is generally lower in amylose content and has a lower gelatinization temperature than does U.S. conventional long grain rice. After cooking, medium grain rice is soft, moist, and sticky in texture. This type of rice is generally preferred by consumers in Japan, northern China, North and South Korea, and several countries in the Middle East. Domestically, California medium grain rice is often used in sushi cuisine, cereal, and beer production.

Short grain

Short grain rice is less than 2 times longer than it is wide (table 1.5). In general, the cooking quality, amylose content, and gelatinization

Table 1.5. Permissible range in dimensions and weight among California long, medium, and short grain rice

Type	Form	Length (mm)	Width (mm)	Weight (g/1,000 seeds)
long	brown	7.2–8.0	2.2–2.3	21–25
medium	brown	6.1–6.5	2.7–2.8	24–26
short	brown	5.1–5.7	3.0–3.3	21–28

Source: California Rice Commission 2000.

temperature of short grain rice is similar to that of medium grain rice. Short rice is also used for making sushi.

Specialty Rice

Rice with cooking or processing quality characteristics different from the standard market classes are known as specialty rices. Production and sale of these unique varieties of rice to niche markets provides a means to diversify a rice-only farming system. Specialty varieties of rice generally have unique sensory and cooking traits that are preferred by a select group of customers for special cuisines and specific products. Niche markets are usually relatively small in terms of the volume of rice traded.

Basmati type

Technically, varieties grown only in the Basmati region of India and Pakistan can be labeled as true Basmati rice. Varieties with similar characteristics but grown elsewhere are marketed as a basmati type. "Basmati" means "the fragrant one" in Sanskrit. This type of rice has been grown in the foothills of the Himalayas for hundreds of years. Its nutlike flavor and aroma is enhanced by long-term storage at low moisture content. This rice is unique in that its grains become very long and thin (extreme elongation) after cooking. Basmati types have a moderately firm cooked texture, are dry and not sticky after cooking, and have an aroma often described as being popcornlike. It has an amylose content and gelatinization temperature similar to conventional U.S. long grain rice. Calmati-202 (CT-202) is a long grain California public variety with the unique Basmati-like quality traits.

Premium quality

Premium quality rice is either medium or small grain in terms of grain length, amylose content, and gelatinization temperature, but it exhibits unique flavor, textural, and appearance characteristics. These varieties have quality characteristics traditionally favored in Japan and Korea, as well as in Japanese American and Korean American communities. The properties that distinguish cooked premium quality rice from standard medium grain rice are its high glossiness, well-defined taste and aroma characteristics, and sticky but smooth texture that remains soft after cooling. Examples of premium quality varieties include Calhikari-201 (CH-201), M-401, Akitakomachi, and Koshihikari. The latter two are not California public varieties; they were imported from Japan.

Waxy (glutinous or sweet)

Calmochi-101 (CM-101), a short grain rice, is the predominate variety of sweet rice grown in California. Although typically a short grain, this style of rice can also be long and medium. Milled waxy rice appears opaque, as opposed to nonwaxy rice, which is translucent. Waxy rice has very little amylose, and cooked milled waxy rice is very soft and sticky. Flour made from waxy rice is also used in products such as candy, salad dressings, baked crackers, and snack foods. Waxy rice is commonly eaten in Laos, Vietnam, Korea, Philippines, and Myanmar. It is also used to make mochi, a traditional rice cake eaten during the Japanese New Year.

Risotto type

This rice type originated in Italy and is used for making various kinds of risotto dishes. Risotto types of rice have a chalky center that is responsible for its ability to absorb the flavor of the stock and produce a creamy texture. There are four principle varieties: Arborio, Baldo, Carnaroli, and Vialoine Nano. No public varieties of this type are currently available in California.

Quality Characteristics of Paddy Rice

Paddy rice quality traits affected by weather, production techniques, harvesting, and postharvest handling are

- moisture content
- immature grains
- purity
- varietal purity
- degree of purity
- yellowing
- scalped or cracked grains
- discolored grains
- surface fissuring

Moisture Content

The proper moisture content (MC) is important at harvest and during milling. Harvesting too soon results in immature grains, and harvesting too late subjects dry kernels to rehydration fissuring from wet-dry cycles caused by dew or rain. Moisture content and temperature during the drying process are also critical as they determine whether small fissures or full cracks develop. The optimal moisture content to harvest medium and short grain varieties is from 18 to 24 percent, depending on the variety. California long grain varieties should be harvested at around 18 percent moisture content. Dried rice is typically milled between 13 and 14 percent.

Immature Grains

Immature rice kernels are small and chalky. Small kernels can be lost during sorting, and chalky kernels result in excessive production of bran and broken grains during milling.

Immature rice kernel with chalky, white endosperm. Photo: Michael Poe.

Varietal Purity

A mixture of varieties in a field may substantially diminish the value of the crop due to downgrading and reduced milling yield. Mixtures may result in low initial husking efficiencies, nonuniform whitening, and lower grading of milled rice.

Saving seed from a production field that has not been managed as a seed field is a potential source of mixed varieties. The use of certified seed greatly reduces the chance of mixed varieties.

Degree of Purity

The degree of purity is related to material present other than rice, including rice straw, rocks, weed seeds, and soil. These impurities, referred to as dockage, generally come from field operations. Dockage in the grain may increase drying costs and reduce milling yield.

Yellowed rice. Photo: Michael Poe.

Yellowing

Yellowing is a result of microbiological and chemical activity in overheated grain when paddy rice is exposed to wet conditions. This can occur in the field or when wet rice is not aerated within several hours after harvest. Long-term storage at or above the recommended moisture content and temperature also causes the endosperm to become yellow to amber. Field mold damage in particular is increased by unfavorable weather conditions or lodged rice in an undrained field. The presence of fermented or yellowed grain does not affect milling yield, but it does downgrade the appearance of the milled rice. Insect- or mold-damaged grains may have black spots around the germ end of the brown rice kernel. Discoloration of the endosperm is only partially removed by milling.

Scalped and Cracked Grains

A delayed harvest results in overexposure of mature paddy rice to fluctuating moisture conditions, leading to fissures and cracks in the individual kernels. Harvesting with combines that are improperly adjusted can also damage individual kernels.

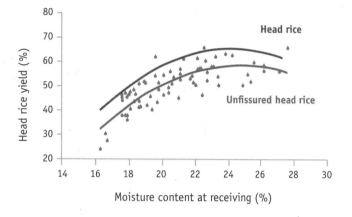

Figure 1.3. Effect of receiving moisture content on surface-fissured head rice (% of whole kernel milled rice) of mixed California medium grain rice varieties *Source*: Mutters and Thompson 2002.

Milled whole rice kernels with fissures. *Photo:* Michael Poe.

Surface Fissuring

Surface fissuring is an increasingly important measure of quality in both the domestic and export specialty rice markets. Surface fissures are small cracks in the surface of milled rice that are too small to qualify the kernel as cracked in a traditional sense. Surface fissures are important because they are an indication of whether a kernel will break during cooking, causing poor eating texture. One way to evaluate the degree of surface fissuring is to soak milled rice in room-temperature water for 30 minutes. Fissured whole kernels tend to break when soaked. A high incidence of broken kernels using this method can significantly reduce the value of the rice. The cause of surface fissures is unclear. It does not appear to be related to harvest moisture content based on paddy rice samples from medium grain varieties taken at a Sacramento Valley dryer receiving station in 2002 (Mutters and Thompson 2002). The percent of kernels with surface fissures remained consistent across a range of harvest moisture contents of 16 to 28 percent (fig. 1.3). Surface fissuring may occur during heated air drying

and may be influenced by the rate of moisture removal from the milled rice during handling.

Discolored Grains

Genetic traits notwithstanding, the color of milled rice may be shades of white, creamy, gray, or rosy as a result of production and handling practices. USDA standards allow color limits to be applied within grades. For example, U.S. No. 1 must be white or creamy, but U.S. No. 3 may be lightly gray. Kernels may be materially colored as a result of heating. Color standards also apply to parboiled kernels.

Physical Properties

General Appearance

The general appearance of the rice is a particularly important trait because most rice is consumed in the whole grain form. Factors associated with general appearance are grain size and shape, uniformity, translucency, chalkiness, color, and damaged and imperfect grains. These are subjective evaluations performed by the human eye, although recent advances have been made in automating the process. General appearance of the milled rice is frequently the first quality rating assigned to a lot of milled rice. U.S. quality grades are based on established criteria (tables 1.3 and 1.4).

Grain Characteristics

Grain type categories are based on the length, width, and weight of the milled kernel. Varieties of each of the three grain types must conform to rather narrow limits of dimension and weight (see table 1.5). These are among the first criteria of rice quality that breeders consider in developing new varieties. If a selection does not conform to recognized standards it is not considered for release. Intensive genetic selection is also done to eliminate undesirable grain traits that detract from or decrease general appearance or decrease milling yields. These defects include furrowed bran, irregularly shaped grains, pointed grain ends, or oversized germs.

Hull and Bran Color

The hull color of U.S. varieties is classified as either light (straw-colored) or dark (gold). In the manufacture of parboiled rice, the hull can color the endosperm in the process and alter the quality of the final milled product. Colored bran varieties are grown for specialty markets and command a premium price. However, the value of a standard Calrose type is reduced if it is contaminated with a colored bran variety. Under the authority of the California Rice Certification Act (Assembly Bill 2622) colored bran types are designated as "commercial impact" varieties. Consequently, the planting and processing protocols are reviewed in advance to ensure that the necessary precautions are taken to prevent any commingling of colored bran types with traditional Calrose varieties during production and processing.

Broken kernels of milled rice. *Photo:* Michael Poe.

Milling Yield

Milling yield has a large impact on the value of rough rice. Two aspects of milling yield frequently measured are head rice yield (HRY), the proportion of milled grains that are greater than 75 percent of the full length; and total rice yield (TRY), whole plus milled broken kernels. Broken rice is generally valued at only 30 to 50 percent of whole grain rice. The actual head rice yield in a milled sample depends on varietal characteristics, production factors, harvesting, drying, the storage moisture content of the sample, and the sample milling method. Accurate measurement of the amounts and classes of whole and broken kernels is done by standardized FGIS procedures. Milled rice is classified into seven types based on length–width ratios: long grain, medium grain, short grain, mixed, second head, screenings, and brewer's.

Milling Degree

The milling degree is a measure of the percentage of the bran removed in the milling process. The milling recovery (proportion of the total rice [TR] and head rice [HR] in a sample) is inversely related to the milling degree; that is, the greater the milling degree, the less milling recovery. Milling degree influences rice color and cooking behavior, which are important aspects of consumer preference. Historically in California, milling appraisals were determined using the western milling procedure. This procedure entailed using a 10-pound weight for the initial milling and a 2-pound weight for polishing with a McGill No. 3 mill. As of 2007, the southern milling procedure was adopted as the standard method for California medium grain and short grain rice. The new procedure uses a 7-pound weight for milling and no weight for polishing. A gain of about 2 percent in head rice yield and a gain of 1 percent in total rice are expected with the southern as compared to the western procedure. The southern procedure also produces slightly less-milled rice, which may result in more surface blemishes in the finished product. Surface blemishes, sometimes referred to as "pecky rice," can reduce the grade.

Translucency and Chalkiness

Essentially all segments of the rice industry require clear, vitreous, translucent endosperms. Consequently, breeding programs routinely incorporate intensive selection for these traits. If part of the milled rice kernel is opaque rather than translucent, it is often described as "chalky." Chalkiness disappears upon cooking and has no effect on taste or aroma, but it is undesirable in virtually all instances. It reduces overall appearance and uniformity, and it generally results in lower milling yields because chalky grains tend to be weak and break easily. An exception to this is parboiled rice, where the process of parboiling tends to strengthen grains. However, even in this application, nonuniformity in chalkiness may result in over- or under-processing some of the grain. Other exceptions where chalkiness is not a detriment to quality are certain Aborio types and sweet or waxy rice.

Whiteness

Whiteness is a combination of varietal physical characteristics and the degree of milling. During

Table 1.6. Range of chemical and physical quality characteristics among California long, medium, and short grain rice

Characteristic	Long grain	Medium grain	Short grain
apparent amylose (%)	22.5–23.3	17.5–18.5	18.5–19.3
ASV*	3.8–4.1	6.0–6.9	6.0–6.2
gelatinization temperature (°C)	73	65–68	68–69
protein (%)	7.3–9.2	5.6–6.9	6.0–6.9
gel consistency (BUI) †	104–133	89–108	95–103
RVA‡			
peak viscosity	203–270	225–290	249–265
hot paste	134–171	120–156	249–265
cool paste	263–345	120–150	244–253
set back viscosity	12 to 74	–40 to 7	–120 to –5
breakdown	95–140	101–139	120–126
pasting temperature (°C)	76.7–77.7	69–71.5	71.3–71.9

Source: California Rice Commission 2000.

Notes:

*ASV = alkali spreading value, 1.5% KOH.
†BUI = Brabender units.
‡RVA = rapid visco-analyzer, AACC method.

the whitening process, the silver skin and the bran layer of the brown rice are removed. Polishing after whitening improves the appearance of the white rice.

Color

Rice color is often referred to as "general appearance." The assessment of this characteristic is typically performed on well-milled, whole kernel rice. The term "whole grain" sometimes refers to unmilled or brown rice. Diseases such as smut *(Tilletia barclayana)* or stack burn *(Alternaria padwickii)* may discolor the endosperm. Also, weedy red rice seeds can give the milled rice a rosy color. Off-colored kernels will not make USDA Grade 1.

Test Weight

The test weight is the weight of a known volume of rice from which relative measures of dockage and unfilled kernels are gathered. Test weight is important in rice storage for estimating the weight of rice in holding bins and the change in weight over time. Test weight as determined by USDA procedures is expressed in pounds per

Winchester bushel. The average weight of U.S. medium grain rice is between 43 and 47 pounds per bushel (1.25 cubic feet).

Physicochemical Properties Affecting Quality

Amylose Content

Starch makes up about 90 percent of the dry matter content of milled rice. Starch is a glucose polymer. The amylose content of starches usually ranges from 17 to 23 percent (table 1.6). Rice with a high amylose content exhibits high-volume expansion (not necessarily elongation) and a high degree of flakiness. High-amylose grains cook dry, are less tender, and become hard upon cooling. In contrast, low-amylose rice cooks moist and sticky. Intermediate-amylose rice is preferred in most rice-growing areas of the world, except where low-amylose japonica varieties are grown.

Based on amylose content, milled rice is classified in amylose groups, as follows:

- waxy (1–2% amylose)
- very low amylose content (2–9% amylose)
- low amylose content (10–20% amylose)
- intermediate amylose content (20–25% amylose)
- high amylose content (25–33% amylose)

The amylose content of milled rice is traditionally determined by using the colorimetric iodine assay index method. The near infrared reflectance (NIR) method for determining apparent amylase is an increasingly common technique. NIR predicts amylose content rather than actually measuring it. Apparent amylose content analysis is widely used for determining rice cooking and processing quality. Apparent amylose content is an indicator of kernel firmness and glossiness of the cooled, cooked rice. It is among the most important quality traits examined during variety development.

Gelatinization Temperature

The time required for cooking milled rice is determined by gelatinization temperature. Gelatinization temperature determines to a large degree the suitability of rice flour as an ingredient in baked and reconstituted foods. Many food processing applications prefer rice with intermediate gelatinization temperature. Environmental conditions such as temperature during ripening influence gelatinization temperature. A high ambient temperature during development results in starch with a higher gelatinization temperature. This character in milled rice is evaluated by determining the alkali spreading value (ASV), a standard assay used to classify the processing and cooking quality of rice cultivars. Rice is classified into high, intermediate, and low gelatinization temperature types based on the extent of grain dispersion when soaked in a potassium hydroxide solution for 16 to 24 hours. Alkali spreading value is inversely related to the temperature at which rice starch granules gelatinize.

Gel Consistency

Gel consistency measures the tendency of the cooked rice to harden after cooling. Within a given amylose group, varieties with a softer gel have a higher degree of tenderness when cooked. Harder gel consistency and firmer cooked rice are associated with high-amylose rice. Hard cooked rice also tends to be less sticky. Gel consistency is determined by heating a small quantity of rice in a dilute alkali solution.

Using a Rapid Visco-Analyzer (RVA) is the standard method for determining rice paste viscosity characteristics listed in table 1.6. These characteristics describe how rice flour behaves during cooking and processing. Amylose content is inversely proportional to the degree of starch granule swelling during cooking and related textural attributes. Generally, rice starch with an amylose content above 22 percent displays low peak viscosity and forms a firm gel upon cooling, as indicated by a high setback viscosity. In contrast, low-amylose rice starch common in medium and short grain varieties has high peak values and low setback viscosity. Starch structure and functionality have a significant impact on the use of cereal grains as food and feed. Starch viscosity characteristics are used to define rice cooking, processing, and eating quality. Many food manufacturers have established criteria for determining the essential quality characteristics for a given application.

Protein Content

Protein content, though less widely used in rice breeding programs, is important because of its direct influence on cooking time, effect on cooked rice texture, and direct impact on nutritional value. Higher protein content causes longer cooking time and firmer texture. High protein content in brown rice is negatively correlated with reduced overall palatability, smoothness, taste, and stickiness in some specialty short grain varieties.

Viscosity

Viscometric properties (hot and cool paste) of the rice paste during heating and cooling are measured as an indicator of the processing characteristics of milled rice and rice flour. These properties indicate how well a particular rice or

rice flour is suited to a production process such as extruded foods or a thickening agent in cooked cereals. Rice with a high pasting temperature and low peak temperature of RVA may impart firmness or hardness to a product such as rice cakes.

Texture Analysis

Consumers worldwide frequently consider texture to be among the most important aspect of cooked rice. Many factors influence texture, including cultivar, physicochemical properties, postharvest handling practices (milling degree, drying conditions, and final moisture), and cooking method. Measuring cooked rice texture attributes is done by sensory and instrumental methods. Sensory analysis by taste panels is sensitive to subtle changes in initial texture perception, including parameters relating to stickiness and adhesiveness. The two-cycle compression test for texture profile parameters measures hardness, cohesiveness, adhesiveness, gumminess, springiness, and chewiness. Increased firmness of rice over time is associated with the starch becoming more crystalline and is referred to as starch retrogradation.

Other Properties

Additional characteristics are sometimes considered in describing rice quality, including analysis of aromatic compounds, lipid content, cooking time, and grain elongation (ratio of cooked to uncooked grain length).

Rice Quality Evaluation in Other Countries

Few universally accepted international standards exist for paddy and milled rice. This is primarily due to differences in the emphasis on the importance of grading paddy and milled rice quality among countries.

Consumers' preferred quality characteristics are not necessarily addressed by USDA rice standards. People from different nations have different taste preferences. In Thailand, people prefer well-milled long grain indica rice that is soft and flaky when cooked. In contrast, Japanese consumers prefer soft, sticky short grained japonica rice. In India and Pakistan fragrant rices bring the highest prices, yet in some Western markets fragrant rice may give the perception of being spoiled. Consumers in Bangladesh, Nigeria, Liberia, and parts of the United States prefer parboiled rice.

Table 1.7. Japanese standards for brown rice

	Minimum values*			Maximum values (%)					
Grade	Weight (g/l)	Perfect grains (%)	Characteristics	Moisture	Damage	Dead	Colored	Off types rice	Foreign matter
1	810	70	See list in note, below	15	15	7	0.1	0.3	0.2
2	790	60		15	20	10	0.3	0.5	0.4
3	770	45		15	30	20	0.7	1.0	0.6
Off grade	< 770	—		15	100	100	5.0	5.0	1.0

Source: Ministry of Agriculture, Forestry, and Fisheries, Tokyo, Japan.

Note: The first 3 categories (weight to characteristics) are minimum acceptable values; the last 6 categories (moisture to foreign matter) are maximum acceptable values. All categories must be met to achieve a specific grade. Characteristics include grain fullness, hardness, size uniformity, shape, luster, skin rub, and chalkiness.

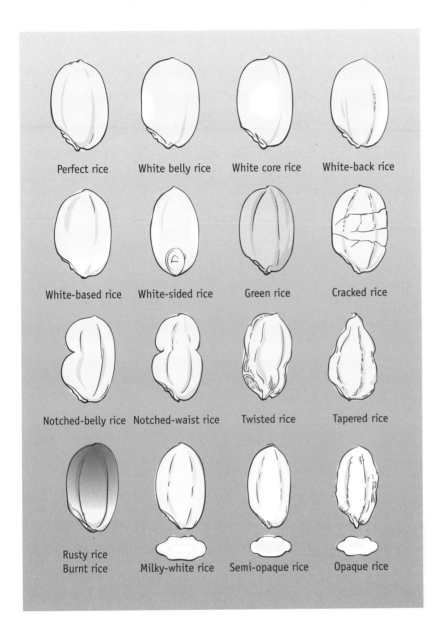

Perfect rice White belly rice White core rice White-back rice

White-based rice White-sided rice Green rice Cracked rice

Notched-belly rice Notched-waist rice Twisted rice Tapered rice

Rusty rice Burnt rice Milky-white rice Semi-opaque rice Opaque rice

The dominate factors describing food processors and consumer preferences in many countries are properties such as whiteness, amount of brokens, shape, chalkiness, amylose content, alkali spreading value (gelatinization temperature), and aroma. Even packaging attributes such size and stated country of origin may influence consumer preference.

Japan

Japan is the largest importer of California rice, so understanding Japanese quality preferences is of interest. In this market aspects of rice quality considered important are appearance of brown and white rice, milling quality, and eating quality. The initial grading of rice in Japan is based on a set of characteristics similar to the USDA system (table 1.7). Components of brown rice such as sound whole kernel, immature kernel, and underdeveloped kernel are important for assessing the quality of rice. An inspector belonging to the Japan Food Agency visually evaluates the components of the rice as an official inspection when brown rice is shipped to the mill.

Gloss and stickiness of cooked rice are also important traits not associated with the appearance of uncooked rice. Good eating quality for Japanese consumers includes high

stickiness, sweet flavor, high gloss of cooked rice, intact whole kernels, and good palatability. The eating quality of a particular lot of rice is often evaluated by a taste panel. Taste panel scores are also used to calibrate "taste analyzer machines" that generate taste scores. In Japan, protein, amylose, and moisture contents are considered to be the primary determinates of taste in white rice. In contrast, brown rice taste scores are a function of protein, amylose, moisture, and free fatty acid content. Most fatty acids are contained in the bran layer, which the milling process removes. Criteria for palpability evaluation used to further grade table rice consist of six sensory tests: appearance, aroma, taste, stickiness, hardness, and overall evaluation.

Rice marketing and import companies frequently have quality standards more rigorous than the national standards established by the Japanese Millers Association. Japanese rice milling companies may impose additional proprietary quality standards that require minimum whiteness standards for brown rice, taste scores, and protein content. Milling appraisals performed according to specific protocols may also be required.

Rice grain grading in Japan: Imperfect rice grains (fig. 1.4)

1. *White belly rice kernels*. Rice with white, opaque areas on the belly or ventral side of the grain. The white area is a result of poorly developed starch granules in the peripheral layers of the endosperm. It is thought to be caused by insufficient transport of carbohydrates to the rice grain during ripening as a result of high temperatures during ripening (the same for white-back, white-based, and white-sided rice). Whitened portions of the endosperm generally do not degrade the quality of the rice.

2. *White core rice kernels*. This condition is more common in large-seeded varieties. The probable cause is excessively high or low temperatures about 15 days after panicle emergence, leading to poor development of starch granules in the central cells of the endosperm. This characteristic is sometimes considered favorable for brewer's rice (i.e., sake) because the fermentation process is more rapid. Sake makers prefer a very high degree of milling in which only 50 to

Figure 1.5. Combinations of fissures that constitute a cracked kernel for rice imported to Japan. *Source:* Redrawn from Matsuo et al. 1975.

| 1 | 2 | 3 | 4 | 5 |

1. Fissure across entire kernel.
2. Fissures extending halfway across kernel starting at the edge.
3. One fissure attached to the edge and another one across three-quarters of the kernel width.
4. Fissure three-quarters of the width of the kernel, regardless of starting point.
5. Intersecting fissures regardless of size.

60 percent of the kernel is recovered after milling, because any remaining fragments of the germ contain a high amount of lipid that impart undesirable flavors to the sake.

3. *Green rice kernels.* This is a result of harvesting and drying while chloroplasts still remain in the pericarp of the grain. White rice with small amounts of green kernels is sometimes preferred by consumers because it affords proof that the rice was not harvested too late and that it is a new crop. However, an excessive amount of green, immature kernels increases protein content, resulting in low taste scores and loss of market value.

4. *Cracked (fissured) rice kernels.* U.S. standards use milling yield to measure the amount of fissured kernels. However, milled whole kernels with fissures tend to break during cooking. Thus, Japanese graders determine the level of whole kernels with surface fissures. Partial fissures and cracks across the entire kernel are considered. They are cumulative: two partial fissures constitute one whole fissure for grading purposes (fig. 1.5).

5. *Notched kernels.* Notched kernels frequently have an hourglass appearance as a result of inadequate grain fill. The bran layer inside the notches remains intact when milled, lowering the grade. If the notches are deep, the kernel typically breaks during milling. The condition is ascribed to low temperatures during early grain fill.

6. *Distorted rice kernels.* Abnormally shaped rice kernels occur when the palea and lemma (hull) do not develop properly. Distorted kernels mill poorly and become rice screenings. These conditions apply to twisted and tapered rice kernels.

7. *Rusty rice kernels.* Brown pigment staining the pericarp is often a result of fungal activity. The grain may be misshapen. Milling does not necessarily remove the off color.

8. *Fermented rice kernels* (not shown). Fermented rice may appear as brown, red, or black markings on the grain surface if exposed to favorable conditions for fungal growth for extended periods of time. The off color may extend into the endosperm. Even short periods of microbial activity may also reduce rice quality. Microbial fermentation can produce off odors and off flavors within a few hours of harvest if the wet rice is not properly aerated.

9. *Milky-white rice kernels.* This condition is when the rice is entirely opaque due to poorly developed starch granules in the center of the endosperm. In general, this rice is slender with nonuniform shape. It is treated as screenings. This condition is most prevalent in the lower kernels of the panicle.

10. *Opaque and semi-opaque rice kernels.* Same as milky-white, except this rice does not have a surface luster. Attributable to unfavorable growing conditions immediately after flowering.

11. *Abortive rice kernels* (not shown). This term refers to the remains of an ovary that stops growing soon after or in the absence of fertilization. Typically referred to as "blanking" in California, this is caused by low-temperature-induced male (pollen) sterility.

Japanese food monitoring program

In 2006, Japan's Ministry of Health, Labor, and Welfare (MHLW) released the Monitoring Plan for Imported Food. The new regulations, referred to as the "positive list," affect all foods imported to Japan after May 2006. The plan outlines the total number of samples to be taken per food group, sample handling and storage, and the chemical tests to be performed. Minimum-access imported rice and tariffed rice are subject to inspection for contaminates such as cadmium, aflatoxins, and agricultural chemicals. The regulation lists over 400 agricultural chemicals that foods may be tested for.

China

New quality grading standards implemented under law GBT 17891 established five classes of rice. Growers are paid according to the quality rather than production weight alone. First class requires a head and total yield percentages of 66 and 70, respectively, in addition to standards for textural characteristics. The higher quality standards are a reflection of China's increased economic affluence and interest in producing rice for export, particularly to Japan.

Taiwan

The population of Taiwan prefers short grain and medium grain rice with relatively low amylose content, low gelatinization temperature, low gel consistency, and translucent grains. Similar to Japan, fissure evaluation is very stringent. Milled rice must have a light reflectance value (whiteness) of 27 or greater and less than 30 percent residual embryo.

Korea

In this emerging market for California rice, the preferred varieties have good translucency, low to intermediate amylose content, low gelatinization temperature, and low gel consistency. High value is placed on low fissuring (internal and surface) scores. Korea has stringent phytosanitary requirements for imported rice. The Food Sanitation Act issued by the Ministry of Health and Welfare is the legal basis for the Korean Food and Drug Administration to establish maximum residue limits on imported foods, including rice. As of July 2004, there are maximum residue levels established for 347 pesticides used on 157 crops (including grain, fruit, etc.). In accordance with the guidelines for inspection, imported crops are subject to testing for 47 pesticides using a "simultaneous multi-residue test."

Philippines

Eating quality is the primary consideration for consumers in the Philippines and is a major concern to rice breeding programs. Consumers usually prefer nonwaxy, medium grain and long grain varieties with clear or translucent kernels and intermediate levels of amylose that remain fluffy and nonsticky after cooking. International Rice Research Institute (IRRI) studies indicate that the major criteria for selecting raw milled rice were grain whiteness and hardness, as well as aroma and flavor for boiled rice. Preference rankings indicate a partiality to white rice with low amylose content. The most desirable waxy varieties are opaque and have bold kernels that are sticky and soft when cooked. Selection criteria for one set of traditional waxy rices are aroma and whole, big grains for raw rice, and aroma and cohesiveness for rice cakes. Instrumental methods for translucency and whiteness of raw rice and hardness and stickiness of cooked rice cannot detect differences perceived by the Filipino consumer panels.

European Union

Spain

Spanish regulations classify rough rice in categories according to variety. Rice is evaluated according to odor, moisture content, foreign matter, insects, defective grains, and milling yield. Spanish consumers prefer short and medium grain varieties and care little about translucency. Rice that cooks as separate and nonsticky grains is preferred, although widely prepared specialty dishes require rice ranging from dry and flaky to soft and creamy. Eating quality is not a consideration in determining selling price.

Germany

German consumers prefer rice that cooks easily and is flaky with a relatively hard texture. Many of the desired qualities for East Asian consumers, such as kernel length and degree of milling, are not important. Price is often of greater interest than quality. In general, rice sold in Germany has a lower percentage of head rice, higher amylose, and harder gel consistency than rice found in many Asian markets.

Table 1.8. Amylose content (AC) type and other properties preferred by rice processors

Rice-based product	W	L	I	H	Other properties/comments
parboiled rice	+	+	++	++	
precooked/quick cooking	+	+	+	+	based on table rice AC
expanded rice	+	+	+	+	AC not major factor
expanded cake, molded		+	++	+	waxy, does not expand
rice cereals and snacks		+	+	+	low fat, texture affected by AC
extruded rice food		+	+	+	low fat
rice-based infant food		+	+		low fat
rice flour and starch	+	+	+	+	wet milling process, freshly milled rice
rice crackers/biscuits	++	++	+	+	*Arare, senbei*
rice pudding	+	+			*Uiro*
rice breads		+	+	+	low gelatinization temperature
unleavened rice bread				++	Pakistani *roti*
rice cakes, steamed	+	+	+		fermented or nonfermented
rice cakes, baked	+	+	+	+	celiac disease
flat noodles/rice paper		+	+	++	low shear process
extruded rice noodles			+	++	hard gel consistency
frozen rice sauces	+	+			gel stability
rice desserts and sweets	+	+			gel stability
fermented rice foods		+	+	++	parboiled *idli, dosai*
rice wines	+	++			low protein, low fat
beer adjunct		+			low gelatinization temperature, low fat
rice in batters and fried foods	+	+	++	+	
rice in thickeners	+	+			gel stability

Key: W = waxy; L = low; I = intermediate; H = high. ++ is preferred more often than +.
Source: Juliano 1998.

Italy

Italy, the largest producer of rice in Europe, imports very little rice. The rice grown there is japonica varieties, which are preferred in Southern Europe. Italian consumer preferences are different from those of Northern Europe, in that they prefer chalky varieties suitable for risotto with intermediate amylose contents and low gelatinization temperatures.

Rice Quality Characteristics Required by Processors

Processors typically require specific grain characteristics to produce a food product with the desired qualities. Examples of principal rice quality traits preferred for selected products are provided in table 1.8. The list is not exhaustive and does not include proprietary specifications employed by selected manufacturers. The overview is intended to underscore the varied concept of rice quality needed for rice products.

Dry Breakfast Cereals

Aged medium and short grain rice with low amylose and a low gelatinization temperature is typically used in U.S. cereal production (see table 1.9). Puffing requires highly milled rice with a moisture content of 13 percent and a lipid content of less than 1 percent. Long grain varieties are unsuitable because of irregular shapes once puffed. For example, Kellogg's Special K cereal uses translucent, highly milled grains with a lipid content of less than 0.5 percent. Breakfast cereal manufacturers are the main distributors of processed rice products, accounting for about one-third of the processed rice market in the United States.

Infant Cereals

Reconstituted infant cereals are a major use of rice flour in the United States. Consumer acceptance demands a cereal that easily reconstitutes with milk or formula to produce a uniform and smooth-textured product. The properties of the rehydrated cereal depend on the hydrolysis characteristics of the starch during processing. Proprietary processes developed by infant cereal manufacturers can accept any rice variety. However, the gelatinization properties of the rice must be well defined in order to establish optimal processing conditions. These properties are associated with the starch fraction of the rice flour. Significant differences exist between the gelatinization temperature of long grain rice and medium and short grain varieties. Typically the starch in long grain rice will gelatinize only above 158°F. In comparison, medium and short grain rice starch gelatinizes below 152°F. This is important because an enzyme (barley malt alpha-amylase) used to control the reconstitution properties of the infant cereal is inactivated at 158°F. Thus, if long grain rice starch is used, the cereal must be heated to 158°F, which would inactivate the enzyme and change the reconstituted properties of the infant cereal. In this application, the quality of the rice (flour) is determined principally by gelatinization temperature alone.

Brewing Rice

Rice is used as an adjunct to barley in the fermentation process of beer production. After corn, rice is the second-most-widely-used adjunct material in the United States in the production of lager beers. Adjuncts are unmalted grains used mainly because they provide a cheaper form of carbohydrate for fermentation than is available from malted barley. Brewer's rice is traditionally the broken kernels from the milling process, although there is a trend for brewers to use whole kernel milled rice. Rice has a neutral aroma and very mild flavor, which are regarded as positive characteristics since the rice will not interfere with the basic malt character of the beer. Rice does, however, promote dry and crisp flavors, a quality favored by many brewers. While rice is preferred over corn grits because of its lower oil content, it does require additional cooking because its gelatinization temperature is too high for adequate starch breakdown during normal mashing. Because of the high gelatinization temperature, not all varieties of rice are acceptable for brewing due in part to high gelatinization temperature and viscosity problems in the cooker. Brewing rice should have carbohydrates that are easily gelatinized and converted into sugars. The growing and storage conditions influence the chemical composition and cooking characteristics of rice. For example, the short and medium grain Calrose types, when grown in California, are satisfactory for brewing requirements. This is not necessarily the case when they are grown in the southern United States. The quality of brewer's rice is determined by several factors, including purity, gelatinization temperature, mash viscosity, mash aroma, moisture, oil content, and protein content. Rice damaged by insects or mold, indicative of improper handling and storage, is unacceptable. Proper storage techniques also eliminate the concern for rancidity, a source of off flavor, which is caused by the oxidation of lipids in the rice. As an example, the rice qualities preferred by the Coors Brewing Company, which has recently increased its use of corn, are listed in table 1.9.

Summary

The definition of rice quality depends on consumer preferences and the intended end use. Specific measured characteristics used to define quality may depend on the consumer. Rice quality is not solely a varietal characteristic; it also depends on the production environment, harvest operations, processing, handling, and storage. All aspects of rice production influence the ultimate quality of the product.

Table 1.9. Representative analysis of preferred brewing rice

Characteristic	Analysis
moisture	11%
total nitrogen	0.8%
soluble nitrogen	0.03%
lipids	0.2%
ash	0.5%
crude fiber	0.5%
gelatinization temperature	61° to 78°C

Source: Coors 1976.

References

Boriss, H. 2006. Commodity profile: Rice. University of California, Davis, Agricultural Issues Center Web site, http://aic.ucdavis.edu/profiles/rice-2006 .pdf.

California Rice Commission. 2000. Variety averages for U.S. market and quality type. California Rice Research Board Web site, http://www.carrb.com/ Datasheets/Variety_Averages.pdf.

Calpe, C. 2007. Review of the rice market situation in 2007. United Nations Food and Agriculture Organization Trade and Markets Division. FAO Web site, ftp://ftp.fao.org/docrep/fao/010/a1410t/ a1410t01.pdf.

Coors, J. 1976. Practical experiences with different adjuncts. Technical Quarterly of the Master Brewers Association of America 13:117–119.

Del Mundo, A. M., D. A. Kosco, B. O. Juliano, J. J. H. Siscar, C. M. Perez. 1989. Sensory and instrumental evaluation of texture of cooked and raw milled rice with similar starch properties. Journal of Texture Studies 20(1): 97–110.

FAS (U.S. Department of Agriculture Foreign Agricultural Service). 2007. Grain: World markets and trends. Foreign Agriculture Service Circular Series FG 01-07. FAS Web site, http://www.fas.usda.gov/grain/ circular/2007/01-07/graintoc.htm.

FGIS (U.S. Department of Agriculture Federal Grain Inspection Service). 2005. United States standards for rice. Washington, DC: USDA. Revised January 2002, updated July 2005. USDA Web site, http://archive.gipsa.usda.gov/reference-library/standards/ricestandards.pdf.

Jongkaewwattana, S. 1990. A comprehensive study of factors influencing rice (*Orryza sativa* L.) milling quality. PhD dissertation, University of California, Davis.

Juliano, B. O. 1998. Varietal impact on rice quality. Cereal Foods World 43(4): 207.

Luh, B. S. 1980. Rice: Production and utilization. Westport, CT: AVI.

Matsuo, T., K. Kumazawa, R. Ishii, K. Ishihara, and H. Hirata, eds. 1975. Science of the rice plant. Vol. 2, Physiology. Part 6, Grain quality. Tokyo: Ministry of Agriculture, Forestry, and Fisheries.

Ministry of Agriculture, Forestry, and Fisheries, Tokyo, Japan. MAFF Web site, www/maff.go.jp/eindex.html.

Mutters, R. G., and J. F. Thompson. 2002. Geographical and environment factors affecting rice milling quality and yield. Annual report to the California Rice Research Board. Project No. RR02-8.

Naito, S., and T. Ogawa. 1994. A new evaluation system for profiling rice eating quality and its suitability. Japanese Journal of Crop Science 63(4): 569–575.

Nelson, R. 2006. Japan releases English version of 2006 monitoring plan for imported food. GAIN Report. USDA FAS Web site, www.fas.usda.gov/gainfiles/200605/146197865.pdf.

Paggi, M. S., and F. Yamazaki. 2001. The WTO and international trade prospects for rice. University of California Executive Seminar on Agricultural Issues, December 2001. UC Agricultural Issues Center Web site, http://aic.ucdavis.edu/oa/paggi.pdf.

Pan, Z., K. S. P. Amaratunga, and J. F. Thompson. 2006. Relationship between rice milling conditions and milling quality. Transactions of the ASABE 50(4): 1307–1313.

Salum, M. 2006. International rice trade: An analysis. Islamabad: Pakistan Agricultural Research Council.

Sauer, D. B., ed. 1992. Storage of cereal grains and their products. 4th ed. St. Paul: American Association of Cereal Chemists.

Siebenmorgen, T. J., and J. F. Meullenet. 2004. Impact of drying, storage, and milling on rice quality and functionality. In E. T. Champagne, ed. Rice chemistry and technology. St. Paul, MN: American Association of Cereal Chemists.

Unnevehr, L. J., B. Duff, and B.O. Juliano. 1992. Consumer demand for rice grain quality. Los Banos, Philippines: International Rice Research Institute.

Zhou, Z., K. Robards, S. Helliwell, and C. Blanchard. 2002. Composition and functional properties of rice. International Journal of Food Science and Technology 37(8): 849–868.

CHAPTER 2

Air Temperature, Humidity, and Weather

Introduction

Weather conditions, particularly air temperature and humidity, have a major effect on rice quality at harvest as well as in bin drying and storage aeration. The study of the heat and water vapor properties of air is called psychrometrics. This chapter defines the commonly used psychrometric variables; describes the psychrometric chart, a graphical depiction of the variables and their relationships; and describes how these variables affect rice quality.

Common Psychrometric Variables

Air temperature is the temperature recorded by a thermometer placed in the open air and shielded from the sun. It is sometimes called dry bulb temperature after the temperature measured by an uncovered thermometer in a wet and dry bulb psychrometer. Air temperature is measured in Fahrenheit or Celsius degrees.

Humidity ratio is the water vapor content of the air on a dry basis (sometimes called mixing ratio or absolute humidity). It is measured in dimensionless units of mass of water vapor per mass of dry air (lb/lb). Water vapor moves from an area with a high humidity ratio to an area with a low humidity ratio. The potential of air to dry rice is best

described by the difference in the humidity ratio between the drying air and the air in equilibrium with the undried rice, rather than the relative humidity of the drying air.

Relative humidity is the amount of water vapor, measured as water vapor pressure, in the air divided by the maximum amount of water vapor the air could hold if it were saturated. It is expressed as a percentage. Relative humidity is a good measure of conditions that allow mold and bacterial growth.

Wet bulb temperature is temperature measured by a glass thermometer covered by a water-soaked cotton wick. In unsaturated air (air with less than 100 percent relative humidity), water evaporates from the wick, cooling the thermometer below the dry bulb temperature of the air. Wet bulb temperature is valuable in describing drying and aeration processes.

Dew point temperature describes the air temperature at which water vapor begins to condense out of the air. For example, as the roof of a metal bin cools at night the air in immediate contact with the undersurface of the roof also cools. When that air cools to the dew point temperature, water vapor begins to form on the underside of the roof. As the roof continues to cool, more water condenses on the roof and can drip onto the rice.

The Psychrometric Chart

The psychrometric chart describes the relationships between the common psychrometric variables. Figure 2.1 is a psychrometric chart in customary units, and figure 2.2 is a chart in metric units.

Air temperature is on the horizontal axis of the chart, and the humidity ratio of the air is on the vertical axis. Under typical California conditions, the humidity ratio of outside air varies between 0.004 and 0.015 lb/lb. Even though water vapor represents only 0.4 to 1.5 percent of the weight of the air, this small amount of water vapor plays a very significant role in rice drying and storage.

The leftmost, upward-curved line in the psychrometric chart gives the maximum amount of water vapor that air can hold at a specific temperature; this curve is called the saturation line. Notice that air holds more water vapor at increasing temperatures. As a rule of thumb, the maximum amount of water vapor air can hold doubles for every 20°F increase in temperature. The saturation line can also be defined as the 100 percent relative humidity line. The 50 percent relative humidity line comprises the points representing the humidity ratio when the air contains one-half of its maximum water vapor content. The other relative humidity lines are formed in a similar manner. Notice that relative humidity alone does not determine a specific air condition on the chart and by itself is not useful in describing the humidity of the air. For instance, 80 percent relative humidity at 32°F has a much lower humidity ratio than air with 80 percent relative humidity at 68°F.

If a mass of air is cooled without changing its moisture content, it loses capacity to hold moisture. If cooled enough, it becomes saturated (has 100 percent relative humidity); if cooled slightly more, it begins to lose water in the form of dew or frost. The temperature at which condensation begins to form is called the dew point temperature if it is above 32°F and the frost point temperature if it is below 32°F. Dew point temperature can be measured directly or determined using the psychrometric chart. For example, temperature and relative humidity can be used to find the state point location of air on the chart; the dew point temperature is the air temperature at the point where a horizontal line drawn from the state point intersects the saturation line. For example, the 87°F air temperature line and the 40% relative humidity line intersect at point A. The dew point temperature of this air is 60°F, point B.

Dew point temperature is directly related to the water vapor content of air. For example, if the dew point temperature of the air is 60°F it has a humidity ratio of 0.011 pound of water vapor

Figure 2.1. Psychrometric chart in customary units.

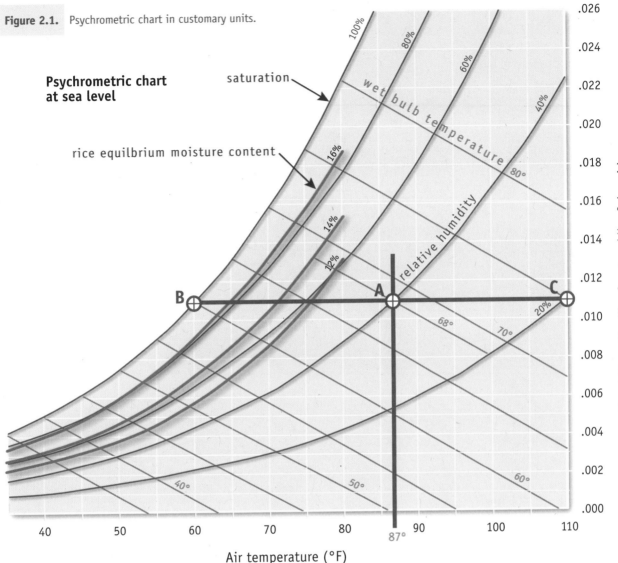

**Psychrometric chart
at sea level**

saturation

rice equilbrium moisture content

per pound of dry air (point C). Dew point is the most commonly available measure of water vapor content. A high dew point temperature represents air with a high water vapor content.

Wet bulb temperature is another commonly used psychrometric variable. On the chart, lines of constant wet bulb temperature slope diagonally upward from right to left. In practice, wet bulb lines are used to determine the state point on the psychrometric chart that represents the air conditions as measured by a psychrometer.

Generally, two psychrometric variables must be measured to determine the complete list

of air properties. For example, the intersection of a diagonal wet bulb temperature line (equal to the temperature of a wet bulb thermometer) and a vertical dry bulb temperature line defines the temperature and humidity conditions of air. On figure 2.1, the 68°F wet bulb temperature line and the 87°F dry bulb temperature line intersect at point A. Once the two variables are used to locate the state point of air on the chart, the other psychrometric values of the air can be determined, as was explained with dew point temperature above.

Psychrometric calculations can also be made with commercial software available

Figure 2.2. Psychrometric chart in metric units.

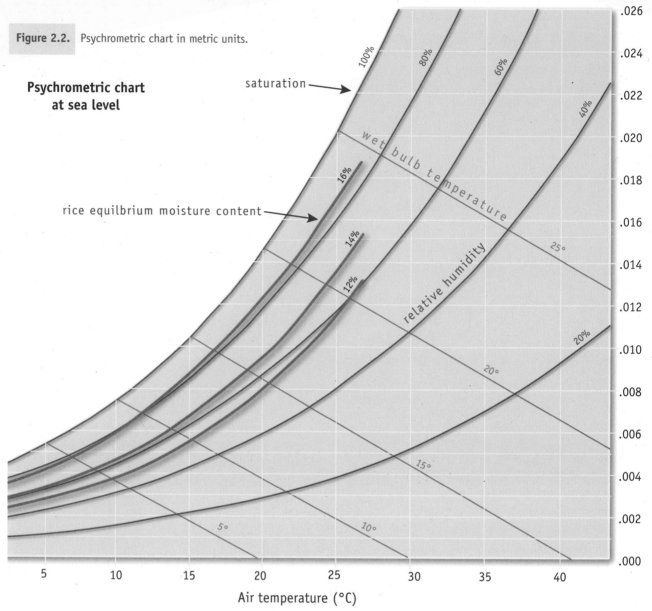

Psychrometric chart
at sea level

saturation

rice equilbrium moisture content

for computers and PDAs. Web sites such as that of the UC Davis Biometeorology Program (http://biomet.ucdavis.edu) can be used to make conversions between measured psychrometric variables and other variables. The site also lists the equations used to make these conversions. Many electronic devices that measure temperature and humidity can display temperature in Celsius and Fahrenheit units as well as several humidity parameters.

Rice placed in an air stream with a constant temperature and relative humidity will equilibrate to a constant moisture content. High relative humidities cause high rice moisture contents, and low relative humidities cause low rice moisture contents. Lines of constant equilibrium moisture content are plotted on the psychrometric chart. They are similar in shape to the relative humidity lines on the chart.

Psychrometric charts and calculators are based on a specific atmospheric pressure, usually at typical sea-level condition. Precise calculations of psychrometric variables require an adjustment for barometric pressures different from that listed on a standard chart. Consult the ASHRAE Handbook listed in the references or the UC Davis Biometeorology Web site for more information on barometric pressure corrections. Most field measurements do not require the accuracy provided by an adjustment for atmospheric pressure.

Aeration, Drying, and Storage Processes on the Psychrometric Chart

Heating air, for example with a direct-fired burner in a column dryer, is represented on the psychrometric chart as a horizontal line extending from the state point of the air to the final air temperature to the right of the state point. The line is horizontal because the air gains heat, causing its temperature to rise, but it does not lose or gain moisture. (Actually the burning gas adds a very small amount of water vapor to the air, but that can be ignored in most calculations.) See figure 2.3 and question 4 in the section "Psychrometric Calculations," below.

Air cooling, as occurs when the sun goes down on a still day, is also represented by a horizontal line on the psychrometric chart, but it extends to the left of the state point. As the air cools, its position on the chart moves to the left at a constant humidity ratio, and it crosses progressively greater relative humidity lines. Eventually it will reach the saturation line and a further decrease in temperature will cause water vapor to condense out of the air. After the air reaches the saturation line, more cooling causes its position on the chart to move downward and to the left along the saturation line. The amount of dew that condenses out of the air can be calculated as the difference between the initial humidity ratio of the air minus the humidity ratio of the air at its coldest temperature. See question 4 in the section "Psychrometric Calculations," below.

Moisture evaporation from rice, as occurs during drying and aeration, follows constant wet bulb temperature lines on the chart (see fig. 2.3). As the dry air moves through the rice, its temperature drops as heat is transferred from the air to the water evaporating from the rice.

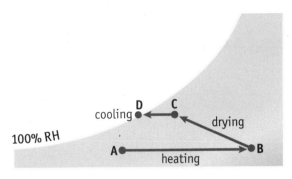

Figure 2.3. Heating, cooling, and drying processes described on a psychrometric chart.

This also causes the water vapor content of the air to increase. The position of the air on the psychrometric chart moves upward and to the left along a wet bulb temperature line as the process continues. The drying stops either when the air reaches the equilibrium moisture content line associated with the moisture content of the rice, or when the air exits the rice. See questions 5, 6, and 7 in the section "Psychrometric Calculations," below.

Psychrometric Calculations

The following questions demonstrate how to use the psychrometric chart to calculate various psychrometric variables and how to predict the temperature and humidity conditions in drying and aeration.

Calculating Psychrometric Variables

1. A wet bulb thermometer reads 60°F and a dry bulb thermometer reads 72°F. What is the relative humidity?

Solution: On figure 2.4 the diagonal 60°F wet bulb temperature (wb) line and the vertical 72°F dry bulb temperature (db) line intersect at point A. Point A is at 50 percent relative humidity (RH), midway between the 60 percent and 40 percent relative humidity lines.

Figure 2.4. Sample psychrometric calculations

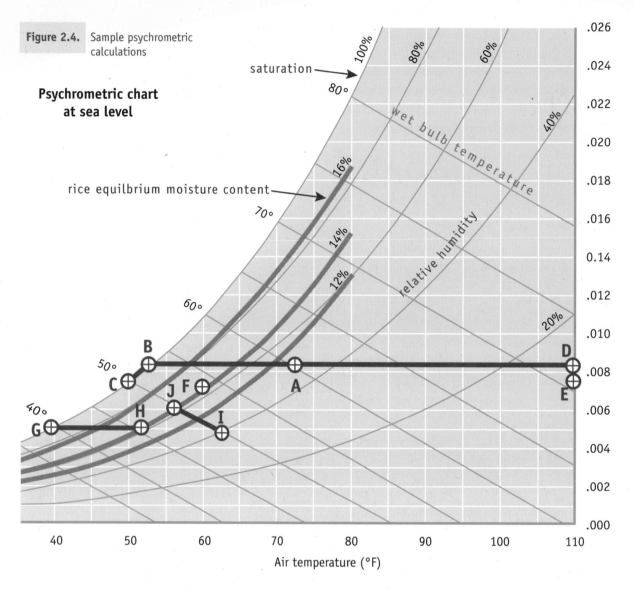

2. What is the dew point temperature of the air in problem 1?

Solution: If the air represented by point A is cooled without changing its moisture content, it will follow a horizontal line until it reaches 53°F, point B. At that temperature, it has 100 percent relative humidity, and any further cooling will cause water to condense out of the air as dew. The dew point temperature is 53°F. If the air cooled to 50°F it would be represented by point C.

3. What is the humidity ratio of the air in problem 1?

Solution: Find the humidity ratio of the air represented by point A by reading horizontally across to the vertical axis of the psychrometric chart. The humidity ratio (w) is 0.0084 lb/lb, point D.

Air Conditions in Drying and Aeration

4. If the air in problem 1 is heated to 110°F, what will be its relative humidity?

Solution: Heating is represented by a line moving horizontally to the right from the initial air condition. The line intersects the vertical 110°F (the maximum on the chart) air temperature line at about 16 percent relative humidity, point D.

This air has a wet bulb temperature of 72°F. If it is used for drying rice in a bin, it will cause moisture to condense on rice if the rice is cooler than 72°F.

On calm, clear nights the air cools by radiation, following a horizontal line to the left. If the air cooled to 52°F (and 100% RH) (point B), it would still result in 16 percent relative humidity if it were heated to 110°F (point D). If it cooled to 50°F, point C, water vapor would condense out of the air, and its humidity ratio would decrease. If this air (even though its relative humidity is 100%) were heated to 110°F it would have less than 16 percent relative humidity (point E). Column dryers with sufficient burner capacity to maintain a constant air temperature throughout the day produce the driest air in the early-morning hours when dew is falling outside.

5. If rice is at 14 percent moisture content and 60°F, could outside air at 40°F and 100 percent relative humidity be used to cool the rice without changing its moisture content?

Solution: Yes, by heating the outside air. The rice is represented by point F on the chart and the air is at point G. If the air were heated to 52°F it would be at point H, on the 14 percent equilibrium moisture content line. Heating would drop the relative humidity enough so that the air would cause no moisture gain or loss from the rice and would still be cooler than the rice, dropping its temperature to 52°F. Another option on a clear day is to postpone aeration until later in the day when the air naturally warms to 52°F. Aeration controllers continuously measure air temperature and relative humidity and automatically operate fans when temperature and relative humidity are optimum.

6. If rice is at 14 percent moisture content and 60°F, and outside air is at 63°F and 40 percent relative humidity, what would be the temperature of the rice if this air is used to completely aerate it?

Solution: The intersection of the vertical 63°F line and the 40 percent relative humidity line falls on the 50°F wet bulb temperature line, point

I. The air passing through the rice will cool and gain moisture following the 50°F wet bulb line upward until it intersects with the 14 percent rice equilibrium moisture line, point J. The air will cool to about 57°F and the rice will also cool to this temperature.

7. Would the aeration cycle described in question 6 result in loss of rice moisture content?

Solution: Yes. The air entering the rice, point I, has a humidity ratio of 0.005 pound of water per pound of dry air. The air leaving the rice, point J, has a humidity ratio of 0.006. The moisture added to the exhaust air comes from the rice and will result in a small amount of drying. The rice on the very bottom of the bin will be at the initial air temperature (63°F) and will dry below 14 percent moisture content.

Psychrometrics and Rice Quality

Humidity, as measured by relative humidity or dew point temperature, and air temperature are critical variables used to control fan operation in dryers and storage facilities. They can also be used to understand the cause of losses in head rice yield. The following generalizations summarize how psychrometric properties change during the day and during drying and aeration.

Psychrometrics and Weather

1. The water vapor content (measured as the humidity ratio or dew point temperature) of outside air usually changes little during a day.

2. Relative humidity in the rice growing area of California changes a great deal during the day because the air temperature changes a great deal during the day.

These relationships are seen in typical temperature and humidity data for the Sacramento Valley (fig. 2.5). Relative humidity drops below 40 percent in the late afternoon hours and peaks near 100 percent in the early-morning hours. It changes because the air temperature varies by nearly 30°F each day. When the air cools, it has less ability to hold water vapor, causing the existing water vapor in the air to represent a high proportion of the total possible water vapor, that is, it has a high relative humidity. As air heats it can hold more water vapor, and the existing air moisture represents less of the total possible water vapor— a low relative humidity. Dew point temperature is relatively constant during the day. During the three days described in figure 2.5, the dew point varies within a range of about 6°F. It is lower in the early-morning hours when water vapor condenses out of the air as dew. It may seem counterintuitive that relative humidity is highest when dew point temperature is lowest, but this demonstrates that relative humidity is not a good measure of the actual water vapor content of air.

3. Dew point temperature is controlled mainly by weather front movement into the region.

During typical calm, clear harvest conditions, dew point temperature remains relatively constant, between 50°F and 60°F. During most harvest seasons there are periods of north wind lasting 2 to 4 days that bring dry air into the rice-growing region and drop dew point temperatures below 40°F (fig. 2.6). When this air warms during the day, its relative humidity can drop below 15 percent. The low relative humidity allows rice to dry to a very low moisture content (the equilibrium moisture content) as the wind causes the rice to lose moisture very quickly. The relative humidity rises at night during these periods, but it is not high enough to allow dew formation, so rice can dry without being subject to rewetting that causes kernel fissuring. During north wind conditions, it is possible to have low drying costs and high head rice yields.

Drying and Aeration Operations

4. Heating air does not change its water vapor content, but it does decrease its relative humidity.

Heating air with an open flame adds virtually no water vapor to the air. This means that the humidity ratio of the air remains constant. However, the relative humidity of the air decreases because the air's ability to hold moisture increases as its temperature rises. The maximum amount of water vapor the air can hold nearly doubles when it is heated by 20°F. This means that the relative humidity decreases by half when the air is heated by 20°F. During heating, air gains temperature and keeps a constant humidity ratio. Adding heat to raise air temperature even 5° to 10°F usually reduces relative humidity enough to allow natural air drying to continue even during periods of prolonged fog and high humidity.

5. Warm air passing through wet rice has a constant wet bulb temperature.

As warm air passes through wet rice it gains water vapor from the drying grain and drops in temperature as it supplies heat to evaporate water in the grain. Air is no longer able to dry rice when its relative humidity equals the equilibrium relative humidity of the rice.

6. As air cools by conduction, it keeps a constant water vapor content (constant humidity ratio) until it cools below its dew point temperature and loses moisture.

When air cools by conduction, as it does when it is next to the cold wall of a metal grain bin in the winter, it drops in temperature but keeps a constant water vapor content. But as the temperature of the air drops, its relative humidity increases (see calculation 4 in the previous section). This means that rice next to a cold bin wall will be exposed to air with a high relative humidity and will gain moisture. If it remains at a high moisture content for long enough it will be subject to the growth of decay. This is why rice next to the interior surfaces of metal

Figure 2.5. Typical air temperature, relative humidity, and dew point conditions for clear, calm days during the harvest season in the Sacramento Valley, California.

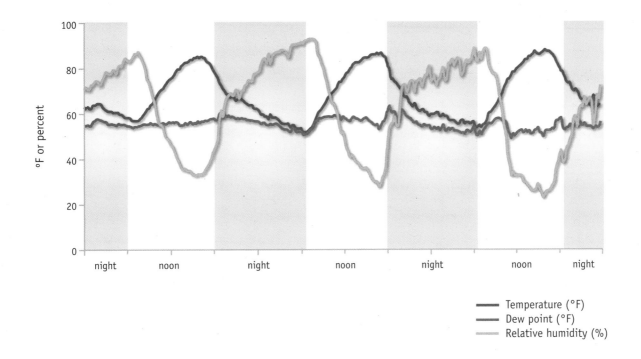

Figure 2.6. Typical change in dew point temperature and relative humidity caused by dry north wind conditions in the Sacramento Valley, California.

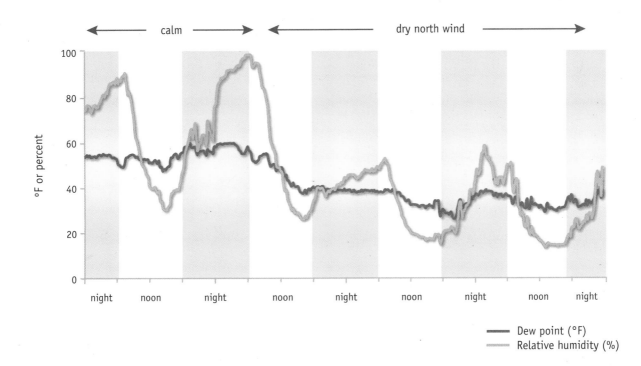

grain bins often becomes moldy. This problem can be prevented by aerating the rice to lower its temperature close to the average outside air temperature.

Measuring Psychrometric Variables

The psychrometric properties of air are determined by measuring two psychrometric variables (three, if barometric pressure is considered). For example, if wet and dry bulb temperatures are measured, then relative humidity, humidity ratio, dew point, and so on can be determined with the aid of a psychrometric chart. The most commonly measured psychrometric variables are dry bulb temperature, wet bulb temperature, dew point temperature, and relative humidity.

Local weather data that includes most of the variables mentioned above can also be obtained on the Internet. For example, the CIMIS system in California (http://wwwcimis.water .ca.gov) has several measurement sites in the Sacramento Valley.

Dry bulb temperature

Dry bulb temperature can be simply and inexpensively measured using an alcohol-in-glass thermometer. The thermometer should be marked in divisions of at least 0.5°F if it is used in conjunction with a wet bulb thermometer. The thermometer should be shielded from radiant heat sources such as the sun, operating motors, lights, external walls, and people. This can be done by placing the thermometer where it cannot "see" the warm object or by protecting it with a radiant heat shield assembly. Glass thermometers respond slowly to changes in temperature. Even with a psychrometer that has a battery-operated fan, accurate readings require several minutes.

Hand-held thermistor, resistance bulb, or thermocouple thermometers can also be used to measure air temperature. Small, low-mass sensors respond to temperature change much faster than glass thermometers. They are more expensive than glass thermometers, but they are not necessarily more accurate. In field situations, a stirred ice-water mixture is an easy way to check calibration at 32°F.

Relative humidity

Electric humidity sensors are based on substances whose electrical properties change as a function of their moisture content. As the humidity of the air around the sensor increases, its moisture increases, affecting the sensor's electrical properties. These devices are more expensive than wet and dry bulb psychrometers, but their accuracy is not as severely affected by incorrect operation. An accuracy of less than 2 percent of the actual humidity is often obtainable. Sensors lose their calibration if they become contaminated, and some lose calibration if water condenses on them. Most sensors have a limited life. The accuracy of the sensor should be periodically checked with a calibration kit that can usually be purchased from the company marketing the sensor. The accuracy of the temperature sensor should be checked using a calibrated thermometer.

Mechanical hygrometers usually employ human hair as a sensing element. Hair changes in length in proportion to the humidity of the air. The response to changes in relative humidity is slow and is not dependable at very high relative humidities. These devices are acceptable as an indicator of a general range of humidity, but they are not suitable for accurate measurements.

Wet bulb temperature

Using a wet bulb thermometer in conjunction with a dry bulb thermometer is a common method for determining the state point on the psychrometric chart. A wet bulb thermometer is an ordinary glass thermometer (although electronic temperature sensing elements can also be used) with a wetted cotton wick secured around the bulb. Air is forced over the wick, causing it to cool to the "wet bulb" temperature.

An accurate wet bulb temperature reading depends on sensitivity and accuracy of the thermometer, maintenance of an adequate air speed past the wick, shielding of the thermometer from radiation, using distilled or deionized water to wet the wick, and using a wick made of cotton.

The thermometer sensitivity required to determine an accurate humidity varies according to the temperature range of the air. More sensitivity is needed at low than at high temperatures. For example, at 100°F a 1°F error in wet bulb temperature reading results in a 1.0 percent error in relative humidity determination, but at 32°F that same error results in a 10.5 percent error in relative humidity. In most cases, absolute calibration of wet and dry bulb thermometers is not as important as ensuring that they read the same at a given temperature. For example, if both thermometers read 1°F low, this will result in less than a 1.3 percent error in relative humidity at dry bulb temperatures between 65° and 32°F. Before wetting the wick of the wet bulb thermometer, operate both thermometers long enough to determine if there is any difference between their readings. If there is a difference, assume that one is correct and adjust the reading of the other by the difference observed before wetting the wick. For example, if the unwetted wet bulb thermometer reads 1° lower than the dry bulb, add 1° to the wet bulb temperature reading when operating the psychrometer.

The rate of evaporation from the wick depends on the air speed past it. A minimum air speed of about 500 feet per minute is required for accurate readings. An air speed much below this will result in an erroneously high wet bulb reading. Wet bulb devices that do not provide a guaranteed airflow cannot be relied on to give an accurate reading.

As with the dry bulb thermometer, sources of radiant heat such as motors, lights, and so on can affect the wet bulb thermometer. The reading must be taken in an area protected from these sources of radiation, or the thermometer must be shielded from radiant energy.

A buildup of salts from impure water or contaminants in the air affects the rate of water evaporation from the wick and results in erroneous data. Distilled or deionized water should be used to moisten the wick, and the wick should be replaced if there is any sign of contamination. The wick material should not have been treated with chemicals such as sizing compounds that affect the water evaporation rate.

Special care must be taken when using a wet bulb thermometer when the wet bulb temperature is near freezing. Most humidity tables and calculators are based on a frozen wick when the wet bulb thermometer reads below 32°F. At temperatures below 32°F, touch the wick with a piece of clean ice or another cold object to induce freezing, because distilled water can be cooled below 32°F without freezing. Below 32°F, the psychrometric chart or calculator must use frost bulb, not wet bulb, temperatures to be accurate with this method.

Under most conditions wet bulb temperature data are not reliable when the relative humidity is below 20 percent or the wet bulb temperature is above 212°F. At low humidities, the wet bulb temperature is much lower than the dry bulb temperature, and it is difficult for the wet bulb thermometer to be cooled completely because of heat transferred by the glass or metal stem. Water boils above 212°F, so wet bulb temperatures above that point cannot be measured with a wet bulb thermometer.

In general, properly designed and operated wet and dry bulb psychrometers can operate with an accuracy of less than 2 percent of the actual relative humidity. Improper operation greatly increases the error.

Dew point indicator

The most common dew point sensor is based on a condensation dew point method. Small amounts of air are passed over a temperature-controlled metal surface. A sensor detects the presence of dew on the surface, and the

temperature of the surface is controlled so dew just begins to form. A condensation dew point sensor is often very accurate over a wide range of dew point temperatures (deviation less than 1°F from −100° to 212°F).

References

ASHRAE (American Society of Heating, Refrigerating and Air-Conditioning Engineers). 2005. ASHRAE handbook, fundamentals. Atlanta: ASHRAE.

Reed, C. R. 2006. Managing stored grain to preserve quality and value. St. Paul: AACC International.

Wexler, A., and W. G. Brombacher. 1951. Methods of measuring humidity and testing hygrometers. National Bureau of Standards Circular 512. Washington, DC: Government Printing Office.

CHAPTER 3

California Rice Commission

Planting and Production

Introduction

Good agronomic practices not only increase grain yields, they also improve the quality of the grain. Variety selection, field maintenance, soil fertility, pest management, irrigation management, and time of harvest all potentially impact quality. Growers must of course consider the trade-offs. For example, using variable drain times to spread harvest may require capital investment to upgrade the existing irrigation system; but compressing the harvest season into fewer days and managing to optimize grain quality requires additional equipment and labor cost and sufficient dryer capacity to accommodate the rapid influx of wet grain.

Variety Selection

Maturity class, grain type, and market demand are important factors when choosing a variety. Information on the time required for maturity of the California public varieties is presented at the end of this chapter. Dimension, shape, weight, volume, and density determine the physical characteristics of rice. In turn, these physical characteristics influence head rice yield (HRY). Short and medium grain rice typically produces higher head rice yields than does long grain rice. This is due to the more rounded, thicker, and harder kernels of the medium grains. Long grain rice is generally less dense than medium grain rice. Additionally, earlier-maturing varieties may yield less head rice than late-maturing varieties, which is also thought to be a result of grain-

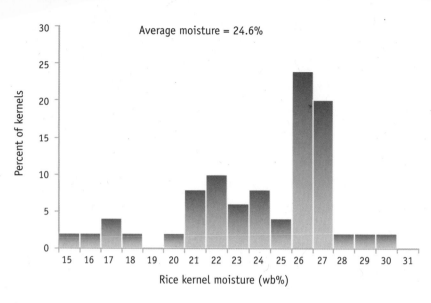

Figure 3.1. Individual kernel moisture content in a sample with an average moisture content of 24%. *Source:* Hill et al. 1998.

filling processes. Evaluation of consumer demand for a particular type of rice is more often done by marketing organizations rather than individual growers. The decision to grow a specialty variety is best predicated on a contract with a distributor. Speculative growing of specialty rice varieties can be a high-risk proposition because of marketing constraints, explicit quality requirements, and the necessary production and processing techniques.

Flowering patterns within the panicle vary somewhat between varieties. Anthesis (flower opening) begins at the top of the panicle and proceeds downward, a characteristic present in all California varieties and referred to as nonsynchronous flowering. The number of days required for all flowers to open ranges from about 4 to 8 depending on the variety. The fewer the number of days required to complete the flowering process, the more uniform the ripening of the panicle. Therefore, the more synchronous varieties may exhibit a greater degree of uniformity of ripening or maturity. The delay in anthesis from top to bottom also means that all flowers do not reach the stage of development that is sensitive to low temperature (e.g., pollen meiosis) at the same time. Brief periods of low temperature result in sections of the panicle being "blank."

Technically, none of the California public varieties exhibit complete synchrony (uniformity) in flowering; however, the duration of flowering within and between panicles on a single plant varies by more than a week, depending on the variety. This range of maturity can be further accentuated by within-field variability in plant growth and development attributable to such things as variable water depth or the uneven application of nitrogen fertilizer. Poor field preparation and management amplifies the inherent genetic variability in ripening. The range of grain moisture content within a field at harvest is a function of management and genetics. The range of moisture content of individual kernels within a panicle can vary from 15 to 30 percent moisture content even though the average may be around 24 percent (fig. 3.1). Research has shown that kernels at 15 percent moisture content or less are likely to fissure when exposed to dew (Thompson and Mutters 2006).

It takes less time to seed a field than to harvest it. Head rice yield in older varieties (e.g. M-202) usually declines as the harvest season progresses, as represented in figure 3.2A, date 1. This decline is attributed to a number of environmental and plant physiological factors. Planting the same variety at two dates could allow for harvesting at higher head rice yields than planting multiple fields on a single date (fig. 3.2A, date 2). Each line in figure 3.2A represents a different field, each planted on a different date. In this example, the expected loss rate in head rice yield remains constant over time. for a given variety. The time difference in planting allows the grower to choose a more opportune time to harvest the individual fields. In a similar fashion, planting varieties from different maturity groups (e.g., M-104 versus M-205) would achieve a similar result. Thus, staggered planting dates or planting rice from several maturity groups enhance the probability of recovering high head rice yield in each field.

The loss of head rice yield over time is much less pronounced in newer varieties. M-205 and M-206 demonstrate a high degree of milling stability throughout the harvest season. A shorter duration of flowering within the panicle in combination with genetic improvements in milling stability has fundamentally changed the relationship between duration of the harvest and head rice yield (fig. 3.2B). The duration of flowering of the traditional and new varieties affects head rice yield over the course of the harvest season. At the early stage of ripening, when the rice is at high moisture content, the head rice yield in M-202 would be slightly better than that of M-206 because M-202's duration of flowering is longer. There will more ripe kernels at the top of the panicle in M-202 than in M-206 at early ripening. As ripening proceeds and optimal harvest moisture is reached, the potential head rice yield of M-206 exceeds that of M-202. The head yield of M-206, unlike M-202, remains relatively constant as the harvest season progresses and the rice moisture content declines.

Field Preparation and Planting

A properly leveled field allows for good management of water depth. Uniform water depth through the course of the season contributes to uniform ripening across the field and consequently more consistent moisture content in the grain between locations. However, cut areas of a field may mature sooner than fill areas due to differences in soil properties attributable to the removal of the topsoil in the cut areas. The affect will likely decrease as the years progress. Operationally the grain produced in the cut areas will dry down sooner and may be more subject to fissuring due to the daily

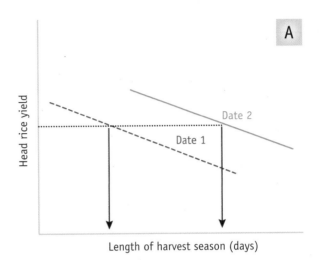

Figure 3.2. Relationship between duration of harvest season and head rice yield under differing planting dates (A) and range of maturity among individual grains (B).

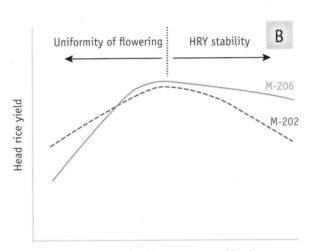

dehydration-rehydration cycles common in the Sacramento Valley during the harvest season.

Level fields contribute to good water management, such as uniform water depth and timely flood-up, which in turn reduces weed competition. Heavy weed infestations can reduce grain quality by outcompeting the rice for resources (nutrients and sunlight) or by contaminating the rice with weed seeds that have high moisture content. Traditionally the

presence of high-moisture weed seeds mixed with dry rice was thought to lead to kernel fissuring; the moisture from the weed seeds rehydrated the rice. This has not, however, been observed under experimental conditions with California medium grain rice.

Unleveled fields have high and low spots. Weeds typically flourish in high spots because the repressive effect of water is absent and because water-active herbicides do not make contact with weeds. Uncontrolled populations of weeds reduce the current year's crop and may reduce the productivity of future crops due to the buildup of the weed seed in the soil. High areas of the field may dry out prematurely during grain ripening. This leads to poor grain fill and an associated loss in milling yields. Low spots in a field not only compromise weed control, they may also negatively impact plant vigor due to delayed stand establishment or increased herbicide stress. The water temperature in puddled areas that remain after the water has been drained for herbicide application can reach 120°F. Such conditions alone or in combination with certain herbicides can impair plant growth. Spatial variability in plant growth and development within a field lead to uneven ripening and inconsistent grain quality at harvest.

Nitrogen Management

Excessive nitrogen leads to delayed grain ripening, late tillering, chalky grains, lodging, and other characteristics that contribute to poor quality. Excessive rates of nitrogen delay overall ripening (fig. 3.3). Experiments conducted on M-201 showed about a 2-day delay in grain ripening for every 25 pounds per acre of nitrogen fertilizer applied (Jongkaewwattana 1990). Delayed harvest results in a higher probability that the rice will be exposed to adverse weather that negatively affects quality. High nitrogen also stimulates late tiller production; late tillers may ripen days after the heads on the main culm and primary tillers. A great many immature

green kernels and a wide range of moisture content in the grain elevates the potential for fissuring and spoilage.

Excessive nitrogen promotes lodging. Rice kernels within a lodged plant canopy have a wide range of moisture content because they do not ripen uniformly. Kernels in contact with or in close proximity to the soil and water are subject to elevated microbial activity and eventual discoloration. Excessive nitrogen also encourages disease that can infect the panicle and impair grain fill.

Protein content, and to a lesser extent amylose content (both intrinsic genetic characteristics), respond to nitrogen management. Protein content may vary by 1 to 2 percent, depending on the management: increasing rates of applied nitrogen cause higher percentages of protein (fig 3.4). High protein content reduces taste quality and rice value in some markets. Amylose content exhibits a similar trend, but its relationship to nitrogen fertilizer application rates and timing is highly variable. In Akitakomachi, the degree to which grain protein content relates to leaf nitrogen levels depends on the rate and timing of nitrogen application.

Recommendations for proper levels of nitrogen fertilization are based primarily on yield response, but these data are a good guide for the levels that produce good quality. UC research has

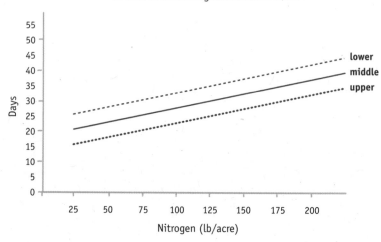

Figure 3.3. Effect of nitrogen on the duration of grain filling in the upper panicle (bottom line), mid-panicle (middle line), and lower panicle (top line) in M-201. *Source:* Jongkaewwattana 1990.

demonstrated that optimal nitrogen rates for maximum yields for California varieties range from 110 to 145 pounds per acre of nitrogen for CH-201 and L-205, respectively (Mutters et al. 2000). Since nitrogen requirements are variable across the Valley, it is important to evaluate these results in context of the individual farming operation. Table 3.1 presents nitrogen requirements of selected varieties as a percentage of M-202 nitrogen requirements. For instance, CH-201 performs well at 75 percent of the optimal rate for M-202; if you traditionally use 150 pounds per acre of nitrogen for M-202, CH-201 would require about 113.

In summary, the newer Calrose varieties exhibit head rice stability across a range of harvest moisture contents, which affords greater flexibility in harvest timing without sacrificing quality. Good agronomic practices preserve the genetically inherent quality potential of all varieties. Uniformity of plant maturation and grain ripening are critical to producing high quality rice. Level fields with consistent water depth throughout, effective weed control, and the correct levels of nitrogen promote uniform ripening and minimize variability in harvest moisture content within a field. Excessive nitrogen delays maturity, encourages lodging, and increases the number of chalky kernels. Short grain varieties are the most sensitive to under- or over-fertilization of nitrogen, followed by medium grain varieties. Long grain varieties are the least sensitive. Harvest moisture and nitrogen management are particularly important to producing high-quality premium short grains.

Water Management

Field Drainage

Well-planned water management during the growing period and timely field drainage prior to harvest are essential to obtain high quality and yield. Too early drainage or lack of adequate water may decrease the number of tillers, increase spikelet sterility and create incomplete grain fill resulting in small or misshapen kernels. Late drainage delays harvest because the soil may not be dry enough to support equipment, preventing harvest at the optimal moisture content and increasing pest and disease pressure should the rice lodge. Although it is not a common practice, midseason field drainage may

Figure 3.4. Protein content in the grain of Koshihikari in response to total nitrogen applied (lb/ac). The individual points represent mean values (n = 4) of nitrogen applied in selected combinations of application times: preplant, preplant/tiller, preplant/PI, and preplant/pollen meiosis. *Source:* Mutters, unpublished data.

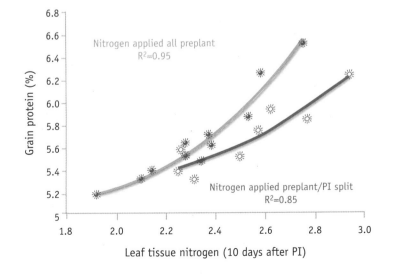

Table 3.1. Nitrogen application rates for twelve California public rice varieties as a percentage of optimal rates of M-202

Variety	N rate as a % of M-202
Short Grain	
CH-201	75
S-102	93
Akitakomachi	60
Koshihikari	50
CM-101	95
Medium Grain	
M-104	104
M-205	102
M-206	96
M-401	96
M-402	95
Long Grain	
L-205	114
CT-201	89

Source: Mutters et al. 2001.

adversely affect yield and grain quality as well. Rice is most sensitive to water stress 5 to 10 days before heading. Inadequate water at heading results in lower seed size and fewer properly filled seeds per head. Nitrogen management does not substantially alter the effect of early or late drainage.

Another important consideration of water management is uniformity of ripeness at harvest. There is an inherent range of grain maturity within a given head because of the way rice flowers. Even though the measured moisture content from a cut sample may suggest an optimal time for harvest, the moisture content of individual kernels may vary widely (see fig. 3.1). Overripened kernels will likely fissure, and green kernels may fail to mature properly yield chalky, misshapen grains. Factors contributing to low-quality, misshapen kernels are presented in chapter 1.

Generalized Production Checklist

Choice of Variety

- Know the quality requirements of the market.
- Know the physical and chemical characteristics of the grain.
- Know the time required for maturity.
- Consider the variety's uniformity of ripening.
- Consider the variety's response to adverse climatic conditions during ripening.

Field Preparation

- Maintain uniform water depth to obtain more uniform grain at harvest.
- Prepare beds and manage weeds properly to reduce weed competition.

Time of Planting

- Consider planting varieties from different maturity groups to stagger harvest.
- Consider grain quality deterioration during ripening (e.g., fissuring versus protein content).

- Match anticipated daily harvest volume with harvesting and drying capacity.

Seeding Rate and Plant Population

- Avoid a high plant population, which reduces yield and quality through mutual shading effects, increased disease incidence, competition for resources, and lodging.
- Avoid a low plant population, which contributes to increased tillering, variation in panicle maturity, weed growth, and low yield.

Nitrogen Fertilization

Excessive rate

- May stimulate late tillering and production of immature panicles.
- Increases male sterility (blanking).
- Decreases number of filled grains.
- Increases lodging.
- Delays harvest.

Inadequate rate

- Reduces grain size.*
- Reduces protein content.
- Reduces lodging.

*Grain size is a genetic trait that is fairly insensitive to management practices. However, inadequate nitrogen can result in poor grain fill and undersized kernels that can be lost during the cleaning process and are therefore not readily apparent in the finished grain.

Source, Rate, and Timing of Pesticides

- Herbicides may modify plant development, tiller production, and prune roots.
- Late applications of pesticide may leave unacceptable levels of residue in grain.
- Timeliness of pest control may influence quality.

Environmental Conditions during Grain Fill and Maturation

- High temperatures accelerate ripening.
- Low humidity and desiccating winds cause rapid drying.

Characteristics of Selected Rice Varieties Grown in California

Short Grain Varieties

S-102

Grain type: Short
Maturity: Very early
Vigor (poor = 1; excellent = 5): 4.3
Lodging (%): 19
Blanking tolerance relative to A-301 (worst) and M-103 (best) (worst = 1 [A-301], best = 5 [M-103]): 4
Stem rot score (worst = 10, best = 1): 7.1
Height: 36.2 in.
Relative yield potential (lb/ac), 1996: 96% of M-202
Days to 50% heading: 81

Comments: Very high yield potential and 2 weeks earlier than S-201. Good resistance to low-temperature blanking. Grain is 8% larger than S-201, with less chalkiness. Rough leaves and hulls; grain dries down rapidly during ripening. Susceptible to stem rot.

S-201

Grain type: Short
Maturity: Early
Vigor (poor = 1, excellent = 5): 4.6
Lodging (%): 27
Blanking tolerance (worst = 1 [A-301], best = 5 [M-103]): 3
Stem rot score (worst = 10, best = 1): 6.2
Height: 35.3 in.
Relative yield potential (lb/ac, last 5-yr avg.): 92% of M-202
Days to 50% heading: 96

Comments: High yield potential. Excellent seedling vigor. Similar to M-201 in maturity and resistance to blanking. Good pearl shape, larger seed size than other short grains.

Calhikari-201

Grain type: Short
Maturity: Early
Vigor (poor = 1, excellent = 5): 4.5
Lodging (%): 40
Blanking tolerance (worst = 1 [A-301], best = 5 [M-103]): 3
Stem rot score (worst = 10, best = 1): 6.0
Height: 34.6 in.
Relative yield potential (lb/ac, last 5-yr avg.): 8,000
Days to 50% heading: 90

Comments: Greater yield potential and lodging resistance than Akitakomachi and Koshihikari. High milling potential but somewhat less than the Japanese varieties. Average taste scores slightly less than Japanese varieties, although marketing organizations indicate that Calhikari-201 is in the premium category.

Calmochi-101

Grain type: Specialty
Maturity: Very early
Vigor (poor = 1, excellent = 5): 4.2
Lodging (%): 60
Blanking tolerance (worst = 1 [A-301], best = 5 [M-103]): 5
Stem rot score (worst = 10, best = 1): 6.3
Height: 36.5 in.
Relative yield potential (lb/ac, last 7-yr avg.): 94% of M-202
Days to 50% heading: 89

Comments: A sweet, glutinous rice. Matures 2 weeks earlier than S-201. Excellent resistance to low-temperature blanking. Has rough leaves and hulls, no awns. Grains dry down rapidly during ripening. Be careful not to contaminate with other varieties. ASCS nonprogram rice.

Medium Grain Varieties

M-103

Grain type: Medium
Maturity: Very early
Vigor (poor = 1, excellent = 5): 4.0
Lodging (none = 0, all = 99): 51
Blanking tolerance (worst = 1 [A-301], best = 5 [M-103]): 5
Stem rot score (worst = 10, best = 1): 5.4
Height: 36.2 in.
Relative yield potential (lb/ac, last 5-yr avg.): 93% of M-202
Days to 50% heading: 90

Comments: Earliest variety, vigor less than M-202. Excellent resistance to blanking. Good head and total milled rice yields. Moderate lodging. Good yield potential, about 7% less than M-202 at normal planting dates. Alternative variety for M-202 in coldest rice-producing areas and for late (or delayed) planting in warmer areas.

M-104

Grain type: Medium
Maturity: Very early
Vigor (poor = 1, excellent = 5): 4.5
Lodging (none = 0, all = 99): 38
Blanking tolerance (worst = 1 [A-301], best = 5 [M-103]): 5
Stem rot score (worst = 10, best = 1): 5
Height: 35.9 in.
Relative yield potential (lb/ac, last 5-yr avg.): 98% of M-202
Days to 50% heading: 92

Comments: Traits similar to M-103 except with a higher yield potential and better early-season vigor, particularly in the cool production zones. An alternative variety for M-202 in coldest rice-producing areas and for late (or delayed) planting in warmer areas.

M-202

Grain type: Medium
Maturity: Early
Vigor (poor = 1, excellent = 5): 4.4
Lodging (%): 22
Blanking tolerance (worst = 1 [A-301], best = 5 [M-103]): 4
Stem rot score (worst = 10, best = 1): 6.4
Height: 36.5 in.
Relative yield potential (lb/ac, last 5-yr avg.): 100% of M-202
Days to 50% heading: 90

Comments: Very high yield potential. Performs better than M-201 in cooler growing areas. Matures 3 days earlier, ripens more uniformly, and more resistant to blanking than M-201. Moderate lodging. Threshes easily but does not shatter. Harvest moisture should not be below 20% or above 22%.

M-204

Grain type: Medium
Maturity: Early
Vigor (poor = 1, excellent = 5): 4.2
Lodging (%): 14
Blanking tolerance (worst = 1 [A-301], best = 5 [M-103]): 4
Stem rot score (worst = 10, best = 1): 6.3
Height: 34.6 in.
Relative yield potential (lb/ac, last 5-yr avg.): 97% of M-202
Days to 50% heading: 93

Comments: Very high yield potential. Seedling vigor lower than M-202, higher than M-201. Height and heading date like M-201; matures very close to M-202. Lodging resistance intermediate between M-201 and M-202. Improved total milling and head rice yield. Resistance to blanking similar to M-202. Threshes easily like M-202. Not recommended for the Escalon area.

M-205

Grain type: Medium
Maturity: Early
Vigor (poor = 1, excellent = 5): 4.2
Lodging (%): 16
Blanking tolerance (worst = 1 [A-301], best = 5 [M-103]): 4.5
Stem rot score (worst = 10, best = 1): 5.4
Height: 36.6 in.
Relative yield potential (lb/ac, last 5-yr avg.): 105% of M-202
Days to 50% heading: 92

Comments: Very high yield potential. Performs better than M-202, particularly in the warmer areas. Not recommended for cooler growing areas. Matures 3 to 5 days later than M-202. Low lodging potential. Threshes easily but does not shatter. Harvest moisture should not be below 18% or above 22%.

M-206

Grain type: Medium
Maturity: Early
Vigor (poor = 1, excellent = 5): 4.5
Lodging (%): 18
Blanking tolerance (worst = 1 [A-301], best = 5 [M-103]): 4.5
Stem rot score (worst = 10, best = 1): 5.4
Height: 37.4 in.

Comments: Very high yield potential. Better-adapted to cooler areas than M-205; when compared to M-202, heads about 5 days earlier, shows improved resistance to blanking, and potentially better head rice yield (3%). During ripening the kernel moisture may "hang" for a few days at around 26%.

M-208

Comments: A blast-resistant variety with yield potential and agronomic characteristics similar to M-202.

M-401

Grain type: Medium, premium quality
Maturity: Late
Vigor (poor = 1, excellent = 5): 4.3
Lodging (1 = none, 99 = all lodged): 30
Blanking tolerance (worst = 1 [A-301], best = 5 [M-103]): 2
Stem rot score (worst = 10, best = 1): 5.6
Height: 38.7 in.
Relative yield potential (lb/ac, last 5-yr avg.): 96% of M-202
Days to 50% heading: 111

Comments: Premium quality rice with large kernels. Good yield potential but susceptible to blanking, lodging, and damage from early drainage. Uses somewhat less nitrogen than other varieties. Best adapted to warmer areas. Milling yields lower than other medium grains.

M-402

Grain type: Medium, premium quality
Maturity: Late
Vigor (poor = 1, excellent = 5): 4.3
Lodging (1 = none, 99 = all lodged): 20
Blanking tolerance (worst = 1 [A-301], best = 5 [M-103]): 2
Stem rot score (worst = 10, best = 1): 2.1
Height: 38.7 in.
Relative yield potential (lb/ac, last 5-yr avg.): 96% of M-202
Days to 50% heading: 105

Comments: Semidwarf, late maturing, smooth, premium quality medium grain. Earlier maturity, improved translucency, and 5% higher head rice yield than M-401. Kernel size and weight is smaller than M-401.

Long Grain Varieties

L-204

Grain type: Long
Maturity: Early
Vigor (poor = 1, excellent = 5): 4.1
Lodging (%): 5
Blanking tolerance (worst = 1 [A-301], best = 5 [M-103]): 3.5
Stem rot score (worst = 10, best = 1): 6.5
Height: 33.5 in.
Relative yield potential (lb/ac): 94% of M-202
Days to 50% heading: 86

Comments: High yield potential. Resistant to lodging. Seedling vigor is fair and may be affected by deep water. Improved head rice yield and cooking characteristics, better than L-202 and L-203. Avoid early draining (requires 40–45 days after 50% heading to mature) and harvest at 18–19% moisture to maximize milling yield.

L-205

Grain type: Semidwarf long
Maturity: Early
Seedling vigor (poor = 1, excellent = 5): 3.9
Lodging (%): 10
Blanking tolerance (worst = 1 [A-301], best = 5 [M-103]): 4
Stem rot score (worst = 10, best = 1): 5.8
Aggregate sheath spot score (worst = 10, best = 1): 2.2
Height: 35.8 in.
Yield (1994–98): 8,608
Days to 50% heading: 88

Comments: An early maturing, smooth, semidwarf long grain rice with improved milling yield and dry cooking characteristics. Heads 2 days later than L-204. Seedling vigor is slightly weaker and about 2 inches taller than L-204. Slightly more susceptible than L-204 to stem rot and aggregate sheathspot. Do not grow in the coolest rice areas of California. Harvest at 16 to 18% for maximum head rice yield. L-205 is suitable for traditional U.S. long grain markets and for processing.

Calmati-202

Grain type: Aromatic long
Maturity: Early
Vigor (poor = 1, excellent = 5): 3.9
Lodging (1 = none, 99 = all lodged): 15
Blanking tolerance (worst = 1 [A-301], best = 5 [M-103]): 4
Stem rot score (worst = 10, best = 1): 5.1
Height: 39.4 in.
Relative yield potential (lb/ac, last 3-yr avg.): 7,032
Days to 50% heading: 91

Comments: Amylose content similar to L-204 and about 2% less than A-201. Kernel size smaller than L-204 or A-201. Produces high milling yields. Suited to the warmer growing areas. Should be harvested at 18 to 19% MC.

Specialty Varieties

Kokuho Rose

Grain type: Medium, specialty variety
Maturity: Late
Vigor (poor = 1, excellent = 5): 4
Lodging (%): 87
Blanking tolerance (worst = 1 [A-301], best = 5 [M-103]): 3
Stem rot score (worst = 10, best = 1): 5.5
Height: 45 in
Relative yield potential (lb/ac, last 5-yr avg.): 74% of M-202
Days to 50% heading: 111

Comments: Proprietary variety for specialty market.

Akitakomachi

Grain type: Short
Maturity: Very early
Vigor (poor = 1, excellent = 5): 3.5
Lodging (%): 83
Blanking tolerance (worst = 1 [A-301], best = 5 [M-103]): 2
Stem rot score (worst = 10, best = 1): 7.0
Height: 39 in.
Relative yield potential (lb/ac, last 5-yr avg.): 70% of M-202
Days to 50% heading: 86

Comments: Premium quality Japanese variety that is agronomically inferior to California rice varieties but has exceptional milling yields. Seedling vigor weaker than most California varieties, about equivalent to A-301. Slow seedling growth may increase risk of herbicide damage and poor stand establishment. Hulls are pubescent, resulting in a quicker drydown than smooth varieties. Straw is weak and lodging is usually severe. Low nitrogen rates should be used, 30 to 60 pounds of nitrogen per acre preplant followed with topdressing at panicle initiation if needed (no more than 100 pounds total). Harvest at higher moisture content.

Koshihikari

Grain type: Short
Maturity: Late
Vigor (poor = 1, excellent = 5): 4.2
Lodging (%): 98
Blanking tolerance (worst = 1 [A-301], best = 5 [M-103]): 2
Stem rot score (worst = 10, best = 1): 7.7
Height: 44 in.
Relative yield potential (lb/ac, last 5-yr avg.): 62% of M-202
Days to 50% heading: 100

Comments: Similar to Akitakomachi except typically requires less nitrogen (40 to 60 pounds total); if following a Calrose fertility program, total nitrogen requirements may be less. A field dedicated to this variety improves performance stability. Sometimes shows a favorable yield response to post-PI topdressing with nitrogen.

References

Champagne, E. T., J. F. Thompson, R. G. Mutters, J. A. Miller, and E. Tann. 2004. Impact of storage of freshly harvested paddy rice on milled white rice flavor. Cereal Chemistry 81(4): 444–449.

Chancellor, W. J., and V. Cervinka. 1973. Harvest improvements with optimum combine management. Transactions of the ASAE 18(1): 59.

Hill, J. E., M. W. Canevari, R. G. Mutters, S. Scardaci, J. F. Williams, and R. L. Wenning. 1998. Cooperative extension rice variety adaptation, and cultural practice research and education program. Annual Report to the California Rice Research Board, Yuba City, CA.

Jongkaewwattana, S. 1990. A comprehensive study of factors influencing rice (*Orryza sativa* L.) milling quality. PhD dissertation, University of California, Davis.

Mutters, R. G., and J. F. Thompson. 1999. Rice quality. Annual report to the California Rice Research Board, Yuba City, CA.

———. 2000. Maintaining rice quality after harvest. Annual Report to the California Rice Research Board, Yuba City, CA.

Mutters, R. G., J. F. Williams, M. W. Canevari, R. L. Wennig, and J. E. Hill. 2000. Cooperative extension rice variety adaptation, and cultural practice research and education program. Annual Report to the California Rice Research Board, Yuba City, CA.

———. 2001. Cooperative extension rice variety adaptation, and cultural practice research and education program. Annual Report to the California Rice Research Board, Yuba City, CA.

Thompson, J. F., and R. G. Mutters. 2006. Effect of weather and rice moisture at harvest on milling quality of California medium grain rice. Transaction of the ASAE 49(2): 435–440.

CHAPTER 4

Randall Mutters

Machine Harvesting and Rice Grain Quality

Introduction

Harvesting affects the yield and milling quality of rice, so growers must harvest rice as carefully as they grow it. Value for most classes of rice in the United States is determined primarily by the percentages of whole milled grain (head rice yield) and total milled grain, although additional quality characteristics related to appearance and eating quality may also be used in certain markets. For example, the 2008 crop USDA loan values for medium and short grain rice averaged $.098 per pound for head rice, and broken rice was worth about $.067 per pound. At a yield of 80 hundredweight per acre, a drop in head rice yield of 1 percent reduced crop value by $2.48 per acre. As the overall market value of rice changes, the value of milling quality changes proportionately. High milling quality pays dividends. During harvesting, what can growers do to maximize milling quality?

Figure 4.1. Effect of drainage timing (DAH = days after heading) on head rice yield over the course of the harvest season for M-202. *Source:* Mutters and Thompson 2007.

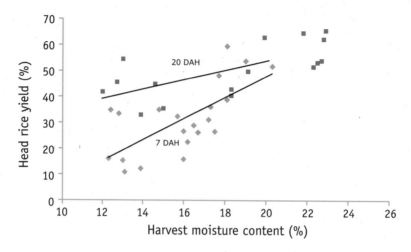

Figure 4.2. Effect of drainage timing (DAH = days after heading) on head rice yield over the course of the harvest season for M-206. *Source:* Mutters and Thompson 2007.

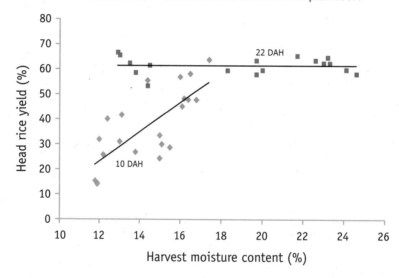

Drain at Optimal Time

The California industry contends that in general draining late produces high head rice yield. A limited number of studies conducted on the high-clay-content soil at the Rice Experiment Station in Biggs, California, confirm that older medium grain varieties like M-202 should be drained about 3 weeks after the crop has reached the 50 percent heading stage (fig. 4.1). Draining at 1 week after 50 percent heading causes M-202 to be more susceptible to loss of head rice quality as harvest moisture decreases below about 21 percent and causes some loss in yield.

In contrast, M-206 exhibited a distinctly different response to drain time (fig. 4.2). Trials conducted on high-clay-content soil demonstrated that the head rice yield of M-206 is less influenced by early draining than M-202. When the field was drained 22 days after 50 percent heading (DAH), the head rice yield remained constant over a range of harvest moisture contents. Draining at 10 DAH resulted in a decline in head rice yield at moisture contents less than 18 percent. M-205 showed greater resistance to fissuring than M-202 but less than M-206. Most important, these results show that for medium grain varieties, the decision to drain a field can be based on the days after 50 percent heading. These new varieties afford growers a greater degree of control over late-season water management without adversely affecting quality. Yields for all tested varieties were unaffected by the drain dates later than 10 to 14 days after 50 percent heading.

Harvest at the Optimal Moisture Content

Many studies have shown that head rice quality is highest when rice is harvested at a moisture content higher than that which is required for safe storage. The optimal harvest moisture content differs with variety; recommended harvest moisture contents are described for some varieties in table 4.1. As a general rule, the older medium grain and short grain varieties should be harvested between 20 and 23 percent, and the long grain and aromatic varieties should be harvested at moisture contents between 16 and 20 percent. There is good evidence that the new medium grain varieties M-205 and M-206 resist fissuring and have good head rice yield even when their harvest moisture content is within a range of 18 to 20 percent.

Any field of rice at harvest time is a mixture of wet and dry kernels, due to growth of the crop, cultural practices, and genetics (see chapter 3, "Planting and Production"). If the dry kernels in such a mix of harvested rice are below approximately 15 percent

Goals of Harvesting

- Harvest at the right moisture.

- Head rice quality declines when rice moisture content drops below a minimum:

  * 20 to 23% for short grain and older medium grain varieties

  * 18 to 20% for M-205 and M-206

  * 16 to 20% for aromatic and long grain varieties

- Sample for rice moisture near noon for consistent results.

- Based on harvest capacity and acreage, start the harvest when rice is slightly above the minimum moisture content to ensure that the last field is harvested at an acceptable moisture content for good quality.

- Harvesting at high moisture results in high drying costs.

- During conditions with no dew at night, rice can be harvested below the minimum moisture content and still have good head rice yield (HRY).

- Plan to increase harvest capacity during periods of north wind.

- Minimize time between harvest and the start of drying to prevent development of off flavors.

- Adjust and maintain harvester to minimize yield loss and grain fissuring.

Table 4.1. Suggested generalized management of selected California rice varieties

Variety	Comment
Short Grain	
CM-101	Dries rapidly during ripening; uniform maturity; harvest at 20–23%.
S-102	Watch carefully, pubescent variety that loses moisture rapidly; has high head potential; harvest at 20–22%.
Calhikari-201	High milling yield potential but somewhat less than Japanese varieties. Quality assessment for this and other Japanese-type varieties goes beyond milling yields; harvest at 22–24%.
Akitakomachi	High milling yield potential; harvest at 23–24% moisture to minimize fissuring.
Koshihikari	High milling yield potential; harvest at 23–24% moisture to minimize fissuring.
Medium Grain	
M-104	Very good milling quality; yield and quality may be reduced by early planting in warm areas; head rice yield stable over a wide range of harvest moisture, but total increases slightly at lower moisture; harvest above 20–22%.
M-202	Ripens uniformly, threshes easily; head rice yield is sensitive to declining moisture; harvest above 20–23%.
M-205	Slightly higher head rice and total yield potential than M-202 when harvested at similar moisture content; cool weather delays heading more than M-202, which could affect milling quality; harvest at 18–20%.
M-206	Five days shorter than M-202 with slightly better yield potential and comparable milling quality; harvest at 18–20%.
M-401	Generally lower milling yields than other medium grains; head rice yield sensitive to low-moisture harvest; harvest above 20–22%.
M-402	Average 5% higher head rice yield than M-401; harvest above 20–22%.
Long Grain	
L-204	Harvest at 16–18% moisture; avoid early draining; requires 40–45 days after heading to mature. Less resistant to cold weather blanking than medium grains.
L-205	Harvest at 16–18% moisture but avoid letting moisture drop too low before harvest; don't begin harvest until grain moisture is below 19% and green grains on the panicles are less than 1%.
A-201	Poor milling yield; use slower cylinder speed; harvest at 18–20%.
Calmati-202	Suggest handling similar to L-205; harvest at 17–19%.

moisture content are sufficiently rehydrated from any source of moisture, they can crack. When these cracked kernels are milled, they usually break, lowering the head rice yield. When average moisture content drops below 21 percent, some dry kernels are subject to fissuring from rehydration. This relationship of moisture content at harvest to milling yield gives growers some control over head rice yield by allowing them to time their harvest according to moisture content. However, while moisture content at harvest is very important, weather conditions and the history of the rice moisture content also determine milling quality, so it is possible to get high or low head rice yield over a wide range of harvest moisture.

Environmental Effects on Head Rice Yield

Rice harvested at low moisture content often does not produce low head rice quality if it has not been exposed to rehydrating conditions. During the dry north wind periods that commonly occur during harvest, rice can dry to quite low moisture contents and maintain high head rice quality because dry conditions prevent nighttime dew; therefore the rice is not subject to rehydration. However, when the north wind ceases and dew-forming conditions return, head rice yield drops. In weather conditions with high dew point temperatures, rice can rehydrate to fairly high moisture contents, levels that are normally associated with high head rice quality (fig. 4.3). Rice that rehydrates after a north wind can produce poor head rice quality even though it is harvested at recommended moisture contents. The history of rice moisture content is an important aspect of understanding the head rice yield produced in a particular field. Soil type also influences the time course of head rice loss. For example, a more rapid decline in head rice yield would be expected on light-textured soils exposed to dry, windy conditions.

Dry, windy conditions cause rapid reductions in rice moisture. For example, in 2003 and 2004 at the Rice Experiment Station, the harvest moisture content dropped 6.2 and 8.2 percentage points by the end of the windy period (fig. 4.3; table 4.2). During the north wind, head rice yield declined from 63.8 to 55.6 percent in 2003 and from 58.2 to 49.7 percent in 2004. However, the grower's return per acre decreased by only $0.08/cwt in 2003 and $0.17/cwt in 2004. During the dry weather, reduced drying costs offset most of the head rice yield loss. After the wind stopped and the dew returned, the rice gained moisture. A few days following the

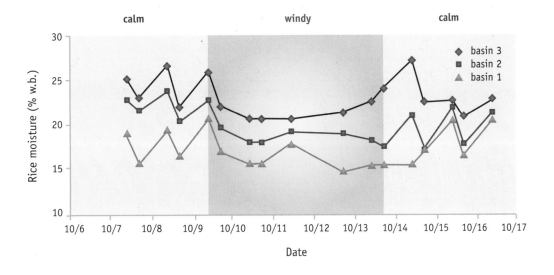

Figure 4.3. Diurnal fluctuation in rice grain moisture before, during, and after a north wind period. *Source:* Thompson and Mutters 2006.

Figure 4.4. Head rice yield as related to harvest moisture content before and after a north wind. Data collected in 2003 and 2004 is for M-202 in Biggs, CA. The range of 1999 data collected from the receiving station in West Sacramento, CA, is represented by the shaded area. *Source:* Thompson and Mutters 2006.

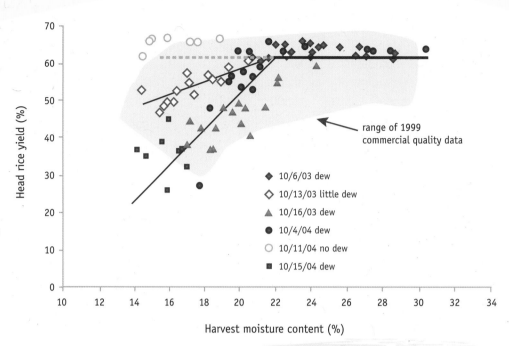

north wind, head rice yield dropped by 9.8 percentage points in 2003 and 24.4 percentage points in 2004. Grower return decreased by $0.54 in 2003 and by $1.25 in 2004 after a few days of rehydration. Complete head rice yield loss due to rehydration takes more than 11 hours of exposure to greater than 90% relative humidity.

Drying costs are a significant expense to the grower. As a rule of thumb a 1 percentage point decrease in harvest moisture content decreases drying costs enough to offset the loss of 1.5 percentage points of head rice yield, based on typical loan values for medium grain rice and typical drying costs. This reduces the economic loss and the risk associated with harvesting at moisture contents that may cause some loss in head rice yield.

Typically, the west side of the Sacramento Valley experiences more north wind days than areas on the east side (fig. 4.5). The number of windy days during harvest ranges from a low of 1.0 around Nicolaus to a high of 6.1 in southern Tehama County. A greater degree of vigilance when scheduling harvests is required in the windy areas.

Environmental Effects on Rice: Points to Remember

- Exposure to the north wind causes little loss in head rice yield.

- Dew after north wind cracks rice. Prevent low-moisture rice from rewetting.

- Prepare for periods of north wind.

- Keep harvest moisture high by spreading harvest dates:

  * Match planting schedule with harvest capacity.

  * Consider using several varieties: M-206, M-202, M-205.

- Harvest as much as possible during the north wind:

  * Harvest 24 hours per day if possible.

  * Keep receiving facilities open late.

  * Provide temporary storage for rice under 18 percent moisture content to free up dryer capacity for handling high-moisture rice.

Rice is transported from the field to the dryer in hopper-bottomed trailers. *Photo:* Randall Mutters.

Sampling for Harvest Moisture Content

Growers should collect a sample prior to harvest to check its moisture content. For best accuracy, use a harvester to cut the sample since a combine sample best represents the true moisture content. Alternatively, one can hand-strip heads from random areas, but be sure to take some of the sample from lower, less-mature panicles. Avoid taking just the ripe grains from the topmost panicles.

On days with heavy dew, a morning sample may be 5 percentage points higher in moisture content than a sample collected from the same location in midafternoon. Collect moisture samples at a consistent time during the day. A meaningful moisture history of a field can be made only when samples are collected at a consistent time after dew has dried from the rice. The harvest moisture content recommendations in this chapter are based on noontime sampling.

Avoid Delays during Harvest and from Field to Dryer

Start harvesting each day when the dew is mostly off the rice to avoid false moisture readings and to avoid putting wet rice in the truck. Stop harvesting when significant moisture accumulates on the plants in the evening and the plants become "tough."

Once harvest has started, complete a field as soon as possible unless significant areas have a higher moisture content; in such cases,

Figure 4.5. Average number of north wind days at selected locations in the Sacramento Valley. Data based on 10-year averages; 1993 to 2002 CIMIS data are from stations located near referenced towns.

Table 4.2. Rice quality and value before, during, and after a dry north wind period in 2003 and 2004 for M-202, Biggs, CA

Harvest date		Moisture content (%)	Head rice yield (%)	Grower return ($/cwt)
2003	Sept. 30	25.9	63.2	5.48
	Oct. 6	24.3	63.8	5.63
	Oct. 13*	18.1	55.6	5.55
	Oct. 16	19.6	45.8	5.01
2004	Sept. 27	22.3	55.8	5.27
	Oct. 4	22.8	58.2	5.46
	Oct. 11*	14.6	49.7	5.29
	Oct. 15	14.3	25.3	4.04

Source: Thompson and Mutters 2005.
Note: *North wind period.

Gerber 6.1
Orland 3.6
Durham 2.7
Colusa 2.4
Nicolaus 1.0
Zamora 4.9
Davis 4.4
Sacramento
Manteca 1.4

leave them until they are ready. Because stripper harvesters move so fast, they should be able to harvest all the grain in a field at closer to the optimal moisture content as compared to cutter bar combines. Over the long run, this should lend itself to higher milling quality.

To reduce quality loss, expedite handling after harvest and throughout the drying process. Rice above 24 percent moisture content begins to produce ethanol within hours after harvest, an indicator of microbial activity. The production of ethanol corresponds to the development of off flavors and odors in cooked rice and is designated as a sour or silage odor by a USDA taste panel. Drier rice is still subject to off odor development, but damage takes progressively more time as moisture content decreases (see chapter 5, "Heated Air Drying," for more details).

Harvest to Maximize Profit

Harvesting is a compromise between loss of grain yield, reduction in quality, and maximizing harvest capacity. Most grain is lost by dropping it on the ground or leaving it on the stalk. Physical damage includes reducing milling quality by cracking, crushing, or shelling kernels. Most adjustments in combine operator's manuals emphasize reducing losses first, which is usually worth more than attempting to improve quality. California studies have found that the average rice field after harvest had about 300 pounds of grain per acre on the ground. This suggests growers

are doing a good job of adjusting combines to minimize losses. After losses are minimized, reduction in grain damage is achieved by adjusting the threshing mechanism. The speed of the combine and the rice variety influence grain recovery performance only slightly when comparing cutter bar headers to stripper headers. Based on header performance studies and southern rice varieties, there were no substantial differences in efficiency between the two header types at typical operational speeds.

Understanding Harvester Basics

A combine is a machine that harvests (cuts and gathers the rice crop) and threshes (separates the grain from the plant and cleans the grain). All combines used in California rice production are self-propelled, which means they have on-board engines that propel the combine and drive the combine mechanisms. Modern rice combines are powerful, with a minimum of 300 horsepower and the capacity to cut a wide swath of 18 to 22 feet. There are several types of combines. Some combines thresh the rice using a cylinder running against a fixed concave that is set crosswise to the travel direction. Axial flow combines have a long rotor that also turns against a fixed surface but it rotates around the longitudinal axis of the combine. Separators include those that use a straw walker to pass the grain along a series of

Figure 4.6. Cutter bar combine, with major systems and functions.

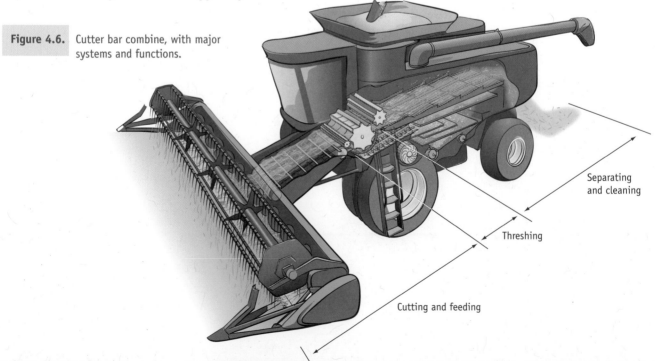

Separating and cleaning

Threshing

Cutting and feeding

ombine equipped with a cutter bar header. *Photo:* Randall Mutters.

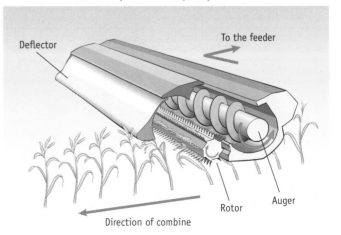

A combine equipped with a stripper header. *Photo:* James Thompson.

"fishback" screens, allowing the grain to fall onto screens and sieves, and tine separators, which employ a secondary cylinder and concave. Axial flow combines use the rear end of the rotor to separate the straw and grain. Figure 4.6 shows the main sections of a cutter bar combine. Cutter bar headers cut the stems and separate the grain and straw internally, while stripper headers strip the grain off the stalk without cutting. Both types are similar in other respects.

Cutter Bar Headers

The parts of a cutter bar header include a reel, a sickle bar, and either a gathering auger or cross draper or both. The reel parts a section of the crop and pushes it against the cutter bar, which cuts the stalk. Rice reels have fingers that help pick up lodged plants. As the cut stalk falls backward into the combine, helped by the reel, the draper helps move it toward the gathering

auger, which in turn carries it to the center of the header into the feeding mechanism (fig. 4.7). Reels can be adjusted forward, backward, up, and down. In standing grain, the bats should strike the grain just below the lowest grain heads. The peripheral reel speed should be 1.25 times greater than the ground speed. The formula for reel speed (RPM) in relation to ground speed is

$$RPM_{reel} = 366 \times 1.25 \times \text{ground speed (mph)} \div \text{reel diameter (inches)}$$

Augers and drapers can be adjusted for clearance from the bottom of the platform and for rotational speed. Auger speed should be sufficient to give a uniform flow of crop into the threshing mechanism. If it is too slow, clogging can occur in the feed area of the platform. Some aftermarket headers use cross drapers, which replace the function of the gathering auger.

Figure 4.7. Cutter bar header, with key components.

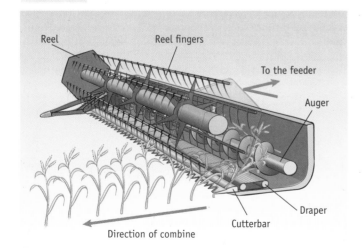

Figure 4.8. Stripper header showing rotor and stripper elements (cover not in place).

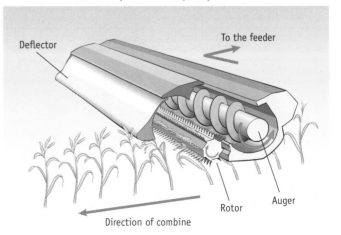

Figure 4.9. Cylinder and concave threshing mechanism.

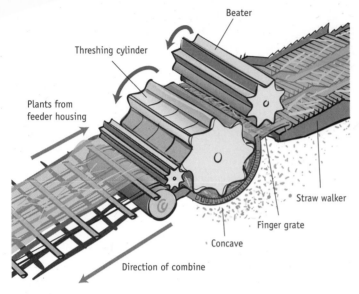

Some operators believe that these cross drapers provide more even feeding.

Stripper Headers

Stripper headers (fig. 4.8) use a rotating drum equipped with stripping fingers that operate under a "cowl" or shroud that deflects the crop under it and briefly traps the heads so the stripping elements can comb off the grain. The rotor rotates upward so that the grain is swept into the header. An auger moves the grain to the feeder mechanism. With stripper headers, the threshing and separation functions of a combine are mainly used to pass the grain on to the cleaning and handling systems. Stripper headers can be attached to standard combines of all types.

Feeding Mechanism

As the cut material enters the throat of the combine, called the feeder house, a feed conveyor feeds it into the threshing mechanism.

Cylinders and Concaves

Cylinders and concaves (fig. 4.9) remove the grain from the stems by abrasive action. Both spike-tooth and rasp bar types are used in rice. The former is more aggressive and may give more complete threshing but with slightly more grain damage, while the latter is considered more gentle but with slightly higher losses. California studies suggest that the virtue of one offsets those of the other.

Axial Flow or Rotary Threshers

Axial flow or rotary threshers (fig. 4.10) can have either single or double rotors mounted longitudinally in the combine. Impeller blades at the front draw the crop in from the

Figure 4.10. Position and orientation of rotor in an axial flow combine.

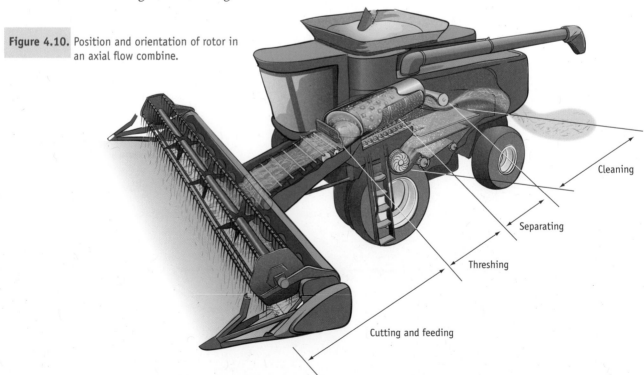

kout wagons move the harvested rice from the combine to transport trailers.
Photo: Randall Mutters.

feeder house. Rasp bars work against concave sections at the front end of the rotor to provide threshing action. The crop is threshed as it swirls between the periphery of the rotor and the concaves.

Separation

Several devices are employed to separate the straw from the grain. In cylinder and concave–type machines a beater located directly behind the cylinder deflects the grain onto the straw walkers, which hold the straw up and allow the grain to fall below while conveying the straw to the rear of the combine. In axial flow machines the rear of the rotor separates and conveys the straw, so there are no straw walkers (fig. 4.11). In all machines the grain then falls through separator grates onto the cleaning mechanism.

Cleaning

To clean the grain, a fan blows air up through the bottom of a chaffer and sieve to separate the chaff and small stems from the grain (see fig. 4.11). The chaffer and sieve together are called the cleaning shoe. The chaffer is an oscillating frame that suspends the straw and chaff so it can be carried out the end of the combine. The sieve, which has smaller openings than the chaffer but operates in a similar manner, does the final cleaning.

Handling

Elevators and augers convey the grain, with clean grain going into the storage tank at the top, or, in the case of unthreshed material from the end of the shoe (called "tailings"), back to pass through the threshing cylinder a second time. From the storage tank an unloading auger moves the grain into bankout wagons driven alongside the combine for transport out of the field.

Adjusting the Combine

Combines must be adjusted to settings unique to the crop, variety, and maturity. Most harvester manuals give initial settings for the crop, but operators must be prepared to adjust these based on conditions in a given field and the threshing characteristics of the rice variety.

Figure 4.11. Separator, cleaning, and handling mechanisms in an axial flow combine.

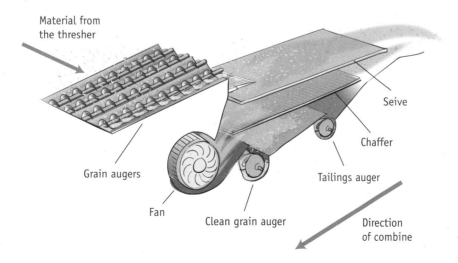

Material from the thresher

Seive

Chaffer

Grain augers

Tailings auger

Fan

Clean grain auger

Direction of combine

Table 4.3. Initial settings for long grain rice varieties recommended by manufacturers (adjust for medium and short grains, see text below)

Operation/Model	JD CTS II	Case/IH 2388	With Shelbourne Reynolds header
rotor speed, rpm	—	800	400–500
cylinder speed	400–550	—	JD CTS: 450–700 Case/IH 2388: 660
concave setting, mm	28	indicator at 1	3–4 mm minimum
fan speed, rpm	1,200	880	full speed
screens	variable–see operator's manual		full open
forward speed	depends on crop ripeness, moisture, standing or lodged, soil moisture, power available, combine capacity, crop yield		

Table 4.4. Effect of increasing cylinder speed by 100 rpm on head rice yield (HRY) and combine associated grain loss

Variety	Change in HRY (%)	Change in combine loss (%)
CalBelle	-0.2	-1.1
M-9	-1.0	-0.5
M-201	-1.0	-0.5
M-401	-2.6	-3.0

Proper Adjustment Prevents Loss and Grain Damage

The initial settings in table 4.3 are reasonable for long grains, which thresh easily. For the more difficult to thresh short and medium grain varieties, California rice growers should increase cylinder speed to 550 to 700 rpm, decrease clearance to 19 millimeters ($^3/_4$ inch), and decrease fan speed to 1,050 to 1,150 rpm. Varieties that are difficult to thresh, such as Koshihikari, might require higher cylinder speed and less clearance.

While harvester settings can affect milling yields, the magnitude is usually less than the effect of rice moisture content unless the settings are far out of adjustment. The settings that are most important to maintaining milling yields are cylinder speed and cylinder clearance, according to studies done in California (Miller et al. 1985). Researchers working on common short stature varieties found a significant effect of cylinder speed on milling yield (table 4.4).

Generally, the economic value of the head rice yield was offset by the increase in recovery as cylinder speed increased. Speeding up the cylinder to 420 to 700 rpm increased revenue for the long grain variety by about 8 percent; but speeding up the cylinder to 500 to 890 rpm did

Table 4.5. Source, cause, and remedies for rice grain losses

Source of loss	Cause of loss	Remedy
preharvest losses	wind and hail shatter, lodging	variety selection and cultural practices
header losses	grain heads missed by cutterbar; grain shattered by reel; grain thrown over in front of the reel	lower cutter bar; reduce reel speed; raise reel, reduce reel speed
threshing losses	unthreshed grain carried over the straw walkers; cracked grain due to overthreshing; cracked grain due to excessive tailings	slow ground speed; slow cylinder, open concave; slow forward speed, increase threshing action
straw walker losses	feeding too much material over the straw walkers	slow ground speed to reduce throughput
shoe losses	grain blown out the rear by high fan speed; too much material on the chaffer	slow fan speed; slow ground speed
leakage losses	open inspection, cleaning, and drainage doors; torn seals, sheet metal holes	close doors; repair seals and holes

Table 4.6. Troubleshooting cutter bar combines, including cylinder and rotor types

Problem	Possible adjustments
excessive free grain in straw	remove rear concave inserts; remove tine grate covers; increase cylinder speed by 30 rpm; increase tine speed (change drive sprocket)
unthreshed grain	close concave; increase cylinder/rotor speed; increase tine speed; add concave fillers
free grain at rear of shoe or under rear axle	change fan speed (could need more or less); open chaffer and/or precleaner; reduce cylinder/rotor speed; open concave; decrease tine speed; reduce ground speed
excessive grain in lower elevator	open sieve, chaffer; reduce fan speed
damaged grain in tank	reduce cylinder/rotor speed; open concave; remove concave filler strips
tank sample is not clean	close precleaner, chaffer, sieve; increase fan speed; decrease rotor speed; open concave
free grain on ground behind header	reduce reel speed
unstripped grain in heads	increase speed of stripper rotor; replace stripping elements; lower rotor; reduce forward speed
excessive shelling of seeds at header	lower header; lower cowl; increase forward speed
trash in sample	close concave; close sieve
cracked grain	reduce cylinder speed; increase forward speed
grain loss over the sieves	increase fan speed if sieves are blocked; decrease fan speed if blowing over; increase forward speed
stripper torque limiter operating excessively	raise header; decrease forward speed; crop too green; increase rotor speed

not change revenue for the medium grain variety M-201, nor did it change revenue for the medium grain variety M-9 between 900 and 1,100 rpm. Cylinder–concave clearance did not make a great deal of difference in this study. In comparison tests of rasp and spike-tooth cylinders, the rasp bar was gentler and produced slightly higher milling yields, but its greater losses offset the head rice yield gain. Both cylinder types produced lower head rice yield and lower losses as cylinder speed increased from a range of 500 to 880 rpm on M-201. Neither had a significant advantage in respect to revenue. Generally, in this study, reducing grain losses was worth more than increasing head rice yield. Each variety required a specific cylinder speed and forward speed for optimal performance.

In more recent work in Arkansas (Andrews et al. 1993), rotor harvester settings were studied for their effects on losses and milling yield. Depending on the variety, the most important factors relating to losses were feed rate (the amount of material in the harvester at any one time, which is affected by forward speed and cutter bar height) and the ratio of material other than grain to grain. Grain moisture content, composition of the material, feed rate, and rotor speed had only small impacts on quality in this study. Field variability had a larger impact than combine settings.

Troubleshooting Combines

Troubleshooting guides for grain loss and damage are usually included in the operator's manual. Identification of problems requires frequent checking for losses on the ground, at the reel, and behind and inside the machine, as well as checking for signs of damaged grain in the tank.

Head rice loss caused by the combine can be estimated by counting the number of grains that have been shelled by the harvester. Each percent of shelled kernels in a lot corresponds to a 1 percent decrease in head rice quality. Tables 4.5 and 4.6, adapted from operator manuals, provide suggested remedies for various problems.

Grain Loss during Harvest

Grain loss due to combine operation may be significant. Typically a well-maintained and adjusted combine leaves about 300 pounds of rice per acre in the field. At the beginning of and periodically throughout the harvest season it is a good idea to measure the grain loss from the combine. See table 4.7 for the number of seeds that reflects 100 pounds of grain per acre when sampled in the straw row for combines with various header and separator widths.

Problem Identification: Measuring Grain Losses

Once the source of an unacceptable loss has been identified, the information in table 4.5 will help you identify the cause and correct it.

Visually

1. Disengage the straw spreader or straw chopper at the rear of the combine.
2. Harvest an area in the field, about 3 to 4 combine lengths.
3. Stop forward motion of the combine and allow the combine to clear itself of material.
4. Back the combine a distance equal to the length of the combine and stop the combine.
5. Walk the area in front of the combine and observe the ground for rice kernels.

NOTE: California rice is generally denser than southern varieties, so the operator manual tables for loss calculations are not accurate. For California medium grain varieties, approximately 35 kernels per square foot equals 100 pounds per acre.

Figure 4.12. Sampling locations for the evaluation of grain loss from harvest operations.

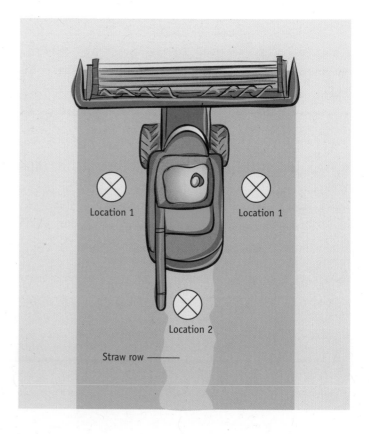

Table 4.7. Loss table for preharvest and thresher losses; values can be adjusted by grain counts per unit area prior to combining to establish thresher loss only

Separator width (in)	Number of rice kernels (medium grain) per square foot to equal 100 lb/ac				
	Cutting width (ft)				
	15	16	18	20	22
29	283	—	—	—	—
38	214	225	188	188	—
44	188	196	225	225	225
55	145	155	283	283	265

By measuring

Begin by following steps 1 to 4 above.

Preharvest losses

1. Place a frame of known area (usually 1 foot square) at several locations in a representative area of the unharvested field.
2. Count the number of kernels on the ground within the frame.

Header losses (with the spreader off)

1. Place a frame (e.g., 1 foot square) at several locations after header (fig. 4.12, location 1).
2. Count the kernels on the ground out of the straw row and take an average.
3. Subtract the average number found in the preharvest test.

Threshing losses (with the spreader off)

1. Place a frame at several locations directly behind the separator area of the combine (fig. 4.12, location 2).
2. Count the number of kernels remaining on partially threshed heads. Average the number of kernels.
3. Count and separately record the number of loose kernels lying on the ground. These will be used to determine straw walker and cleaning shoe losses.

4. Subtract from this result the average number of kernels from the preharvest field loss tests as well as the average number of kernels from the header loss tests.
5. Next, add the partially threshed and loose kernel counts (steps 2 and 3), then use table 4.7 to determine losses attributable to the threshing system.

Straw walker and cleaning shoe losses

1. Determine the average number of loose kernels from threshing losses, above.
2. Use table 4.7 to determine the losses attributable to the straw walker and cleaning shoe.

References

Andrews, S. B., T. J. Siebenmorgen, E. D. Vories, D. H. Loewer, and A. Mauromoustakos. 1993. Effects of combine operating parameters on harvest loss and quality in rice. American Society of Agricultural Engineering 36(6): 1599–1607.

Bennett, K. E., T. J. Siebenmorgen, E. D. Vories, and A. Mauromoustakos. 1993. Rice harvesting performance of Shelbourne Reynolds stripper headers. Arkansas Farm Research 42(4): 4.

Geng, S., J. F. Williams, and J. E. Hill. 1984. Harvest moisture effects on rice milling quality. California Agriculture 38(11): 11–12.

Jongkaewwattana, S. 1990. A comprehensive study of factors influencing rice milling quality. PhD dissertation. Department of Agronomy and Range Science, University of California, Davis.

Kunze, O. R. 1985. Effect of environment and variety on milling qualities of rice. In Rice grain quality and marketing. Los Banos, Philippines: International Rice Research Institute. 37–48.

Miller, G. E., B. M. Jenkins, S. Upadhyaya, and J. Knutson. 1985. Rice harvesting research, In Comprehensive research on rice annual report. Yuba City: California Rice Research Board. 91–98.

Mutters, R. G., and J. F. Thompson. 2004. Factors affecting milling quality and yield. Annual report to the California Rice Research Board, Yuba City, CA.

———. 2007. Crop management and environmental effects on rice milling quality and yield. Annual report to the California Rice Research Board, Yuba City, CA.

Thompson, J. F., and R. G. Mutters. 2005. Crop management and environmental effects on rice milling quality and yield. Annual report to the California Rice Research Board, Yuba City, CA.

———. 2006. Effect of weather and rice moisture at harvest on milling quality of California medium-grain rice. Transactions of the ASAE 49(2): 435–440.

Xu, X. B., and B. S. Vergara. 1986. Morphological changes in rice panicle development: A review of literature. International Rice Research Institute Research Paper Series 117.

Yoshida, S. 1981. Fundamentals of rice crop science. Los Banos, Philippines: International Rice Research Institute.

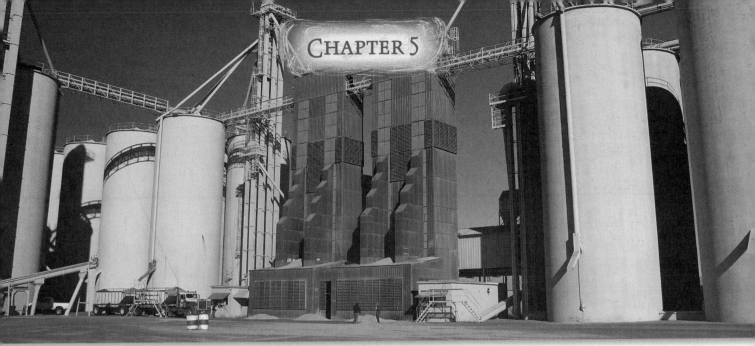

CHAPTER 5

Michael Poe

Heated Air Drying

Keys to Heated Air Drying

Producing high-quality rice with heated air, continuous flow dryers depends on the following steps.

Before Drying

◎ Start drying before rice develops off odors (within 8 hours of harvest if rice has greater than 24 percent moisture content).

◎ Clean rice to remove straw and fines.

During Drying

◎ Dry in several passes through the column dryer, removing 1 to 3 points of moisture per pass.

◎ Obtain best quality, maximum drying capacity, and lowest energy use by operating the dryer at maximum rice throughput rates and highest possible rice temperature (measured at the discharge of the dryer).

◎ Sample rice for quality to determine maximum rice temperature in drying.

◎ Temper rice at least 4 hours after a drying pass, but not long enough to allow microbial activity to cause off odors.

◎ Do not cool rice with ambient air in the dryer.

Limit Time between Harvest and the Start of Drying

Rice should be brought to the dryer as quickly as possible to prevent damage caused by the growth of microbes (bacteria, yeasts, fungi, etc.) and respiration of wet grain. Wet grain held too long may develop off odors, off flavor, and yellow kernels, and may even become contaminated with toxins. The susceptibility of rice to damage is related to moisture content. Rice at less than 21 percent moisture content can be safely held for 2 days before drying begins (fig. 5.1). Rice at greater than 23 percent moisture content should never be held overnight before unloading at the drying facility. Rice harvested at 25 percent moisture content should begin drying within 8 hours after harvest. The safe delay period is the total time from actual removal of the grain from the plant until the rice is exposed to drying air. Rice should not be harvested at moisture contents greater than 25 percent because it is so microbially active and its respiration rate

is so high that it is difficult to start the drying process quickly enough to prevent heating and off odors. Rice at 25 percent moisture content is also expensive to dry (fig. 5.2), and will not have any better head rice yield (HRY) than if it were harvested at lower moisture contents.

Harvesting above 24 percent moisture content requires that harvest and drying operations must be closely coordinated so high-moisture rice is dried shortly after it is received. Wet rice should be aerated and cooled if it must be held for more than 8 hours before drying. Unfortunately there is no foolproof way for a dryer operator to know when a load of wet rice has been stored too long after harvest. Symptoms of a problem load are hot rice in the middle of the trailer and an off odor. Such loads should be dried immediately and may need to be segregated from better-quality rice. The grower should be notified of the problem with the load and should be encouraged or required to delay harvest until rice can dry to a safer moisture content.

Clean Rice Before Drying

Fines consist of weeds seeds, blank grains, hulls, soil, and small pieces of the grain or plant. Newer combine harvesters can separate out most

Figure 5.1. Safe delay period between harvest and the beginning of drying based on average load moisture content, based on data from Champagne et al. 2003 and Matsuo et al. 1995. Safe storage times at moisture contents greater than 20% decrease because commercial loads may have significant amounts of rice at moistures 2 percentage points higher than the average moisture content of the load.

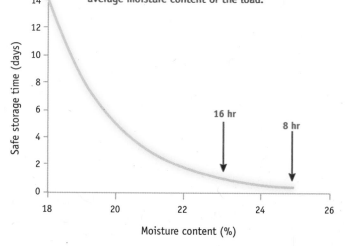

Figure 5.2. Typical costs of drying rice in 2007. *Source: Farmer's Rice Cooperative, Sacramento.*

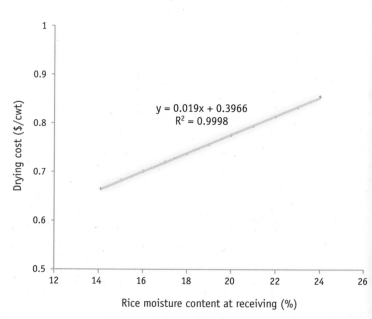

A scalper removes straw and fine materials from the paddy rice.
Photo: James Thompson.

fines during harvest. Any mechanical handling operation, such as column drying or transferring rice, breaks off pieces of hull or grain causing additional fines. Fines reduce airflow and may cause hot spots in storage. Straw that has not been separated from grain by the combine harvester can collect in the dryer, particularly on turn flows (devices used to mix rice in a column dryer), and can block rice flow. Rice with fines should be passed through a cleaner (sometimes called a scalper) before drying. A cleaner uses screens to separate foreign material that is smaller (some weed seeds) or larger (straw) than whole grains and an air blast to separate out low-density material such as blank kernels and hulls.

Dryer Operation

A heated air dryer is operated by controlling the temperature of rice as it exits the dryer and the rate at which rice is discharged from the dryer. The goals are to operate at maximum drying capacity, minimize loss of head rice yield, and minimize energy use.

Loss of head rice yield is minimized by limiting the percentage points of moisture removed per pass. Subjecting a rice kernel to hot air causes differences in moisture content and temperature within the kernel. Increasing differences in moisture content and temperature increase the potential of the kernel to crack. Tempering allows time for moisture content and temperature to equalize within the kernel, so the next drying pass can remove additional moisture without loss in head rice yield. Generally, more moisture can be safely removed per pass when rice is wetter. See the end of this chapter for more details on the drying process.

Dryer capacity is maximized by using its highest possible air temperature. Hot air has a greater potential to transfer heat to rice than does colder air.

Moisture removal is fastest when rice is first exposed to drying air; that is, the first point of moisture is released faster than the second point.

More of the heat added to the drying air is used for drying when rice releases moisture rapidly. When rice loses moisture rapidly, the drying air cools to a lower temperature as it passes through the column of rice than it does when the rice loses moisture at a slow rate.

In practice these principles indicate a dryer operator should

- set the discharge rolls on the dryer to the fastest possible rate

- use the maximum rice discharge temperature that does not cause unacceptable loss of head rice yield

- regularly measure head rice yield to ensure that it is not unacceptably reduced by high discharge rice temperature

The maximum discharge rate will be limited by conveyor capacities and discharge roll speed. If in-feed conveyor capacity limits rice flow rate, set the dryer discharge roll speed so that the receiving bin on top of the dryer is kept full. If the outgoing conveyor capacity limits rice flow rate, set the dryer discharge roll speed so that rice is conveyed without plugging conveyors or spilling at transitions. If conveyor systems do

not limit capacity, the rice discharge rolls can be set at maximum speed.

In typical column dryers operated in California, the minimum discharge rice temperature is about 90°F. Higher temperatures can be used if head rice yield testing indicates minimal quality loss.

Table 5.1 shows the value of operating dryers at high rice discharge rates and high air temperatures. The first series of tests shows that doubling the discharge rate and keeping a constant air temperature increases drying capacity by 8 to 10 percent and increases head rice yield by 2 to 5.2 percentage points, but requires 2 to 3 more drying passes. However, the total time in the dryer for all passes is less when using short residence times than with longer residence times per pass and fewer passes. High discharge rates increase seasonal drying capacity, increase quality, and reduce energy use.

Increasing air temperature from 118° to 136°F and keeping a constant discharge rate causes a 40 percent increase in drying capacity

and reduces the number of passes, but reduces head rice yield by 1.1 to 3.4 percentage points. When both air temperature and discharge rate are increased, drying capacity increases by 37 to 58 percent, and the number of drying passes needed increases by one or two. Head rice quality drops slightly in the column dryer and increases in the LSU dryer.

Dryer Testing

Optimal rice discharge temperature must be determined for each dryer; it may vary depending on the variety and moisture content of the incoming rice. Determine the optimal discharge rice temperature by comparing head rice yield of a lot before it passes through the column dryer with the head rice yield of the rice after it has gone through each drying pass. Collect samples periodically as the lot is dried and then split the composite sample to get a representative sample for a milling test. Continue sampling after each pass through the dryer. Begin

Table 5.1. Effect of different column dryer operating conditions on rice quality and dryer capacity

Test	Dryer*	Discharge rate (cwt/hr)	Air temperature (°F)	Capacity gain[†]	Head rice yield gain	Number of passes
increasing discharge rate	A	960	135	0	0	3
		1,240	136	5%	1.6%	4
		1,760	137	10%	5.2%	5
	B	990	117	0	0	3
		1,910	118	8%	2%	6
increasing air temperature	A	1,260	119	0	0	4
		1,240	136	50%	−1.1%	4[‡]
	B	1,910	118	0	0	6
		1,910	129	12%	−2.9%	5
		1,910	136	26%	−3.4%	4
increasing[§] discharge and increasing air temperature	A	1,260	119	0	0	4
		1,760	137	58%	2.5%	5
	B	960	117	0	0	3
		1,910	136	37%	−1.1	5

Source: Wasserman et al. 1958a, 1958b.
Notes:
*Dryer A is a LSU-type mixing dryer, and dryer B is a cross-flow column dryer.
[†]Capacity gain is based on pounds of water removed per hour.
[‡]High-temperature test removed more total moisture than low-temperature test.
[§]Increasing discharge rate and temperature caused a 15% reduction in fuel use for dryer A and a 3% increase for dryer B.

the testing with a maximum rice discharge rate and a 90°F discharge rice temperature. At this discharge rice temperature there should be little or no loss in head rice yield compared with the initial sample. In the next lot of rice, increase discharge rice temperature by 2° to 5°F by increasing drying air temperature, and then test the quality in the entire batch. If no unacceptable head rice yield loss is observed, the discharge rice temperature can be increased more until a damaging temperature is reached. Many dryers are operated to produce head rice yield losses of less than 1 percentage point for complete drying in the column dryer.

In a typical column dryer system a batch of rice often requires about 8 hours of drying time. Testing may take several weeks because each rice discharge temperature must be tested with the complete range of moisture contents encountered by the dryer. Do not test dryers during periods of wet weather or dry north winds because unusually high or low ambient humidity will cause atypical dryer performance during these periods. Testing should be repeated for different rice types. Japanese varieties may have different drying characteristics than do California-developed varieties. At a constant discharge rice temperature, the dryer will remove less moisture per drying pass as incoming rice moisture content decreases (table 5.2).

Milling tests are done on rice at less than 13.5 percent moisture content. Incompletely dried samples must be dried using a sample dryer. The dryer should be operated with continuous airflow and an air temperature of less than 75°F, or fan cycles on for 2 minutes with a 110°F air temperature and off for 28 minutes. Because there is some unavoidable variation in sample quality evaluations, submit replicate samples for quality evaluation.

Complete dryer testing is needed for newly installed dryers or when an existing dryer is modified. Installation of mixing devices and new controls are examples of modifications requiring dryer testing. New rice cultivars may have different drying characteristics than existing ones, and the dryer should be tested when a new cultivar

Table 5.2. Moisture removed from a lot of medium grain rice dried in a column dryer with a 90°F discharge rice temperature

	Rough rice moisture content	Percentage points of moisture content lost per pass
initial	25.6	—
after pass 1	23.7	1.9
after pass 2	21.8	1.9
after pass 3	19.6	2.2
after pass 4	18.0	1.6
after pass 5	16.3	1.7
after pass 6	15.0	1.3

is introduced. Testing is also needed if regular monitoring, described below, indicates that the dryer is not functioning as it previously did.

Dryer operators should measure and record the incoming and outgoing moisture content of each lot of rice each time it passes through the dryer. Typically, moisture samples are collected every 30 to 60 minutes during an 8-hour pass. When not conducting detailed dryer testing, measure and record incoming and final quality of each lot dried. These data are needed to ensure consistent dryer performance during the season and to provide background data for a diagnosis if damage was done to a lot that later proves to have poor quality.

Tempering

After rice passes through a high-temperature dryer, moisture and temperature within the kernel must be allowed to equalize. Between passes, rice is stored in temporary holding bins, a process called tempering. Maximum head rice yield is obtained with a minimum of 4 to 6 hours of tempering. Tempering also increases the amount of moisture removed in subsequent passes. At moisture contents above 18 to 20 percent, maximum moisture loss is attained after 4 hours of tempering. At lower moisture contents, 12 or more hours of tempering may be required for maximum moisture removal rates. However, removal following 12 hours yields only a small increase in moisture compared with 4 hours of tempering.

Harvest volumes are often so high during the peak of the season that it is necessary to store rice in the tempering bins for more than 24 hours. If the rice has a high moisture content, long tempering times may cause off odors. Completely tempered rice can be aerated to cool it, allowing safe storage. Approximately 1.8 pounds of aeration air per pound of rice is needed for complete cooling. (Note that 1.8 pounds of air equals about 25 cubic feet of air; this air can be supplied over a range of times. See chapter 8, "Storage and Aeration," for more details.) Aeration and complete cooling reduce rice moisture by about 0.5 percentage points. Unfortunately, most tempering bins are quite tall, and high fan power is required to provide the airflow needed for reasonable rates of grain cooling. Rice should not be areated before tempering is completed because this reduces head rice yield.

Types of Continuous Flow Dryers

Column Dryers

Over 80 percent of California's rice is dried in continuous flow, heated air dryers. The most common type is the column dryer (fig. 5.3), in which rice flows by gravity downward between two screens separated by 6 to 12 inches. Heated air flows horizontally through the screens. This dryer is sometimes called a cross-flow dryer because the air flows at a 90° angle to the flow of rice. Discharge rolls (sometimes called metering rolls) at the bottom of the screen section control the rate of rice flow through the dryer. Rice is removed below the metering rolls via screw conveyors. Air is supplied at a rate of 2 to 4 cubic feet per minute-pound and heated to 130° to 165°F, depending on rice residence time in the dryer.

Column dryers have several disadvantages compared with mixing-type continuous flow dryers. Rice tends to flow straight down between the screens, without much horizontal movement. This means that the rice close to the hot air plenum is always exposed to the hottest air and dries more than the rice next to the screen near the air exhaust. Also, rice next to the screens flows slower than the rice in the middle between the two screens. This leads to even more variability in the amount of drying experienced by individual grains. Some column dryers include mixing vanes (sometimes called turn flows) to reduce moisture content variability. Commercial operators have found that finger-type mixers are very effective in reducing kernel-to-kernel moisture variability.

Rice moisture is allowed to equilibrate in temporary storage bins.
Photo: Michael Poe.

Figure 5.3. Schematic of a column dryer. The screen section may be 10 to 15 feet tall in portable dryers and over 70 feet tall in stationary models.

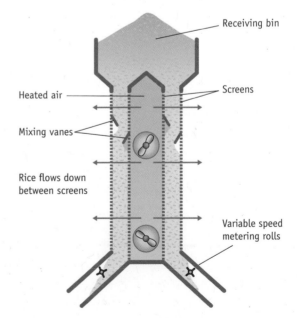

Receiving bin

Heated air

Screens

Mixing vanes

Rice flows down between screens

Variable speed metering rolls

Portable column dryers are available for on-farm column drying.
Photo: James Thompson.

Baffle Dryers and LSU Dryers

Two other types of continuous flow dryers mix the rice as it flows through the dryer. The baffle dryer (fig. 5.4) causes the rice to flow in a zigzag pattern and supplies heated air through openings between baffles. The distance between baffles is generally about 6 inches. The LSU dryer (fig. 5.5) is a large rectangular solid with a series of inverted V-shaped troughs running across the entire width of the dryer. The rice flows

downward past the troughs but does not fill them. The open space below each trough is used to distribute air through the rice. Alternating layers of troughs are air supply or air exhaust channels. This design allows air to take the shortest path from an air supply duct through the grain to an exhaust duct. In some designs the air supply troughs are oriented at a right angle to the air exhaust troughs, and rice mixes as it flows around the air channels. The LSU dryer costs significantly more than a column dryer primarily because of the air pollution equipment needed to control particulate emissions. Because of their higher cost, few of these dryers have been installed in recent years.

Recirculating Batch Dryers

A recirculating batch dryer (fig. 5.6) is the most widely used system in Asia. Paddy rice is loaded into the tempering section and slowly flows downward to the drying section. After passing through the drying section it is returned to the tempering section; the process is repeated until the batch of rice is dry. The units usually remove water at a rate of 0.6 to 1.0 percent per hour and have holding capacities of 1 to 22 tons. The low holding capacities make these dryers unsuitable for the large farming operations in California.

Figure 5.4. Schematic of a baffle dryer.

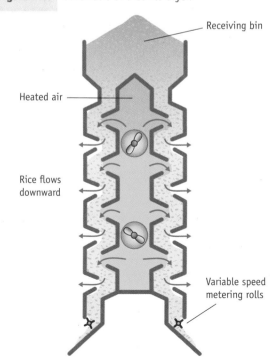

Receiving bin

Heated air

Rice flows downward

Variable speed metering rolls

Figure 5.5. Schematic of LSU mixing dryer.

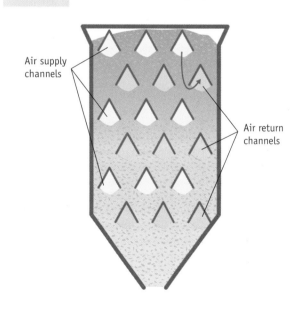

Air supply channels

Air return channels

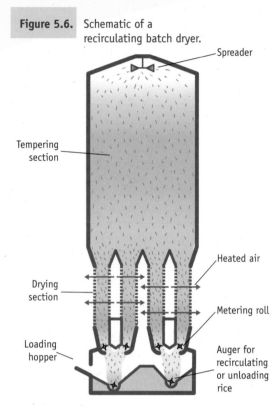

Figure 5.6. Schematic of a recirculating batch dryer.

Labels: Spreader; Tempering section; Drying section; Loading hopper; Heated air; Metering roll; Auger for recirculating or unloading rice

Infrared Dryers

Infrared drying is also used in Asia and is being experimented with in the United States. In this system, paddy rice is passed under a source of radiant heat. Electrical and gas-fired units have been used successfully to provide radiant heat. Preliminary results in California indicate that the units can dry rice quickly but still may require repeated passes, as well as tempering between passes. Some infrared dryers do not use fans to move air through the rice and consequently have reduced electricity costs and may produce less particulate air pollution.

Two-Stage Systems

Rice can be completely dried to milling moisture content in a column dryer or dried to 16 to 18 percent and then finished with unheated air in a flat storage or a perforated-floor grain bin. (See chapter 6, "Bin and Warehouse Drying," for operational details.) This two-stage system works only if the storage has adequate airflow for drying, an airflow greater than that needed for storage aeration. Late-season drying in storage is slow because of low ambient air temperatures.

How Rice Dries in a Column Dryer

In an individual kernel, moisture is lost first from the hull, the part of the kernel directly exposed to the drying air. The endosperm and bran then lose moisture to the hull. In high-temperature drying, the kernel is exposed to drying conditions for about 30 minutes. During this short time the hull may lose 4 to 6 percentage points of moisture content, but the endosperm and bran lose only about 1 percentage point (fig. 5.7). During tempering, the inner parts of the kernel continue to lose moisture and the hull gains moisture. This causes a rather steady loss of moisture from the inner kernel during repeated drying and tempering cycles. The hull goes through cycles of moisture loss and gain.

Increasing air temperature greatly decreases the total residence time in a column dryer (table 5.3 and fig. 5.8). However, removing too much moisture in a short time causes the endosperm to fissure, reducing head rice yield. Acceptable head rice yield can be maintained while using high air temperature by increasing the number of passes through the dryer.

Reducing the amount of moisture removed per pass and increasing the number of passes through the dryer reduces total time in the dryer. Moisture loss is rapid in the first few minutes rice is exposed to drying air because moisture is easily removed from the hull and outside of the kernel. For example, when drying with 160°F air, as described in figure 5.8, rough rice initially

Table 5.3. Effect of drying air temperature and number of passes on total drying time and head rice yield based on drying Colusa 1600 rice from 23.6% to 13.0% moisture content in a 2-inch-deep laboratory dryer

| Air temperature (°F) | 3-pass drying | | 5-pass drying | |
	Time in dryer* (hr)	Head rice yield (%)	Time in dryer (hr)	Head rice yield (%)
100	4.0	62	2.7	65
120	1.8	50	1.3	58
140	1.0	<50	0.7	<50

Source: Adapted from Thompson et al. 1955.
Note: *Total time rice is in the dryer for all passes.*

Figure 5.7. Moisture content of kernel and hull after each drying pass (D) and tempering period (T) in 5-pass drying of rough rice, based on data in Mossman 1986.

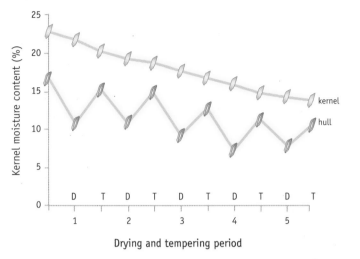

Figure 5.8. Characteristic moisture content loss for a thin layer of kernels of California medium grain rough rice exposed to three air temperatures. *Source:* Adapted from Brooker et al. 1992.

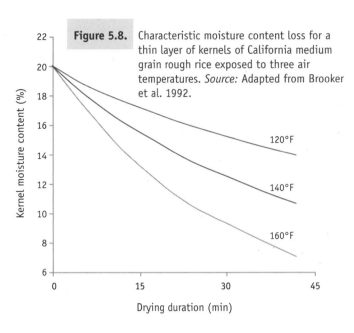

loses moisture at a rate of 0.5 percentage points per minute, but after 30 minutes of drying the rate decreases to about 0.2 percentage points per minute. Moisture deeper in the grain takes time to migrate to the surface; this is accomplished during tempering. Column dryers have more capacity, produce higher head rice quality, and have lower energy use when they are operated with as many passes as possible.

Column dryers have shallow bed depth and very high airflow rates per mass of rice. Adding additional fan capacity rarely increases drying rate.

Drying may cause cracks, or fissures, to develop in the endosperm portion of the kernel. Most kernels with cracks break in milling and contribute to reduced head rice yield. Cracks can be caused by mechanical stress to the kernel, but heated air drying can also cause endosperm cracking. They form after drying and may continue to form for 48 hours after exposure to drying conditions (fig. 5.9). The mechanical properties of the starch change during drying and tempering as a result of the moisture and temperature changes in the endosperm. For example, as the endosperm dries, the loss of moisture in the starch matrix causes it to shrink. Rehydration causes the starch matrix to expand again. Drying causes differences in moisture

content across a kernel and may cause stresses within the endosperm associated with differential contraction of the starch. If the stresses exceed the strength of the starch, the starch will fracture. Starch changes in texture from rubbery to brittle (called glass transition) as it loses moisture and cools. Associated with this transition are also abrupt changes in starch properties, particularly starch matrix density and strength properties. The density change may also give rise to internal stresses and the potential for starch fracturing. The exact relationship between these changes in properties and the mechanisms of moisture transfer through the cell structure of the endosperm are not understood at this time and are the subject of current research. But on the basis of years of experience, we do know

Figure 5.9. Kernel fissuring after 2 hours of exposure to 140°F drying air; Labellle variety with an initial moisture of 21%. *Source:* Adapted from Prasad and Kunze 1982.

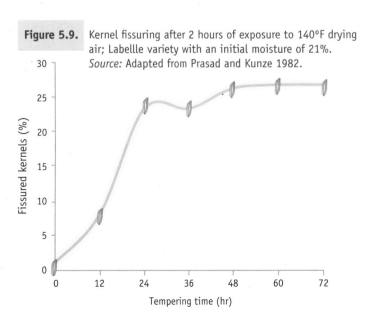

that cracking is minimized by limiting the amount of moisture removed per drying pass. Tempering at high temperature has also been proposed as a method of minimizing cracking.

Energy Use and Conservation

The typical energy use for heated air drying in California is about 3 therms of natural gas and 30 kilowatt-hours of electricity per ton of dried rice. Fuel is used to heat air, and electricity is used to operate the dryer fan, grain conveyors, elevators, and warehouse fans.

Fuel use can be minimized by following the recommendations for increasing dryer capacity and minimizing head rice yield loss. These recommendations reduce the total amount of time the rice is actually exposed to heated air in the dryer, which reduces the energy used for air heating and fan operation. Increasing dryer passes from three to five in the tests described in table 5.1 reduced natural gas use by 3 percent and 15 percent in the two tests.

Air recirculation is a common method of reducing fuel use in many types of convection dryers, but it is not often used in rice dryers. The air exhaust from a dryer has a great deal of fine particles in it, and these particles are a fire hazard if they are exposed to the open flame in direct-fired air heaters.

References

Brooker, D. B., F. W. Bakker-Arkema, and C. W. Hall. 1992. Drying and storage of grains and oilseeds. New York: Van Nostrand Rheinhold.

Champagne, E. T., J. Thompson, K. L. Bett-Garber, R. Mutters, J. A. Miller, and E. Tan. 2004. Impact of storage of freshly harvested paddy rice on milled white rice flavor. Cereal Chemistry 81(4): 444–449.

Cnossen, A. G., and T. J. Siebenmorgen. 2000. The glass transition temperature concept in ice drying and tempering: Effect on milling quality. Transactions of the ASAE 43(6): 61–69.

Frias, J. M., L. Foucat, J. J. Bimbenet, and C. Bonazzi. 2004. Modeling of moisture profiles in paddy rice during drying mapped with magnetic resonance imaging. Chemical Engineering Journal 86:173–178.

Hosokawa, A., ed. 1995. Rice post-harvest technology. Tokyo: Ministry of Agriculture, Forestry and Fisheries.

Kunze, O. R., and D. L. Calderwood. 1985. Rough rice drying. In B. O. Juliano, ed., Rice chemistry and technology. St. Paul: American Association of Cereal Chemists.

Matsuo, T., K. Kumazawa, R. Ishii, K. Ishihara, and H. Hirata, eds. 1975. Science of the rice plant. Vol. 2, Physiology. Part 6, Grain quality. Tokyo: Ministry of Agriculture, Forestry, and Fisheries.

Mossman, A. P. 1986. A review of basic concepts in rice-drying research. Critical Reviews in Food Science and Nutrition 25(1): 49–70.

Prasad, A. D., and O. R. Kunze. 1982. Post-drying fissure developments in rough rice. Transactions of the ASAE 25(2): 465–468.

Steffe, J. F., R. P. Singh, and G. E. Miller. 1980. Harvest, drying, and storage of rough rice. In B. S. Luh, ed., Rice production and utilization. Westport CT: AVI.

Steffe, J. F., R. P. Singh, and A. S. Bakshi. 1979. Influence of tempering time and cooling on rice milling yields and moisture removal. Transactions of the ASAE 22:1214–1218, 1224.

Thompson, J. H., A. H. Brown, W. D. Ramage., R. L. Roberts, and E. B. Kester. 1955. Drying characteristics of western rice—Colusa 1600 variety. The Rice Journal (Oct): 14, 16, 18, 19, 52.

Wasserman, T., M. D. Miller, W. G. Golden. 1965. Heated air drying of California rice in column dryers. California Agricultural Experiment Station Leaflet 184.

Wasserman, T., R. E. Ferrel, V. F. Kaufman, G. S. Smith, E. B. Kester, and J. G. Leathers. 1958a. Improvements in commercial drying of western rice: Part I, Mixing type dryer. The Rice Journal (Apr–May).

———. 1958b. Improvements in commercial dry-ing of western rice: Part II, Non-mixing columnar-type dryer. The Rice Journal (Apr–May).

James Thompson

Bin and Warehouse Drying

Bin Drying

Rice can be dried from harvest moisture content to safe storage moisture content in a storage bin using outside air. About 20 percent of the rice crop in California is dried with this method, in particular, rice produced for seed. It is less expensive than using column dryers because a separate dryer is not needed, and in many seasons drying can be accomplished without gas for air heating. If properly managed, bin drying produces excellent rice quality. The main disadvantage with bin drying is long drying times, but in most years a well-designed and well-operated bin dryer can finish rice drying before the cold months of late fall.

Minimizing Quality Loss in Bin and Warehouse Drying

- Clean rice before putting it in the bin.

- Fill bin to the proper depth based on rice moisture and fan capacity.

- Operate fans continuously until rice reaches desired storage moisture content.

- Use a calibrated moisture meter, humidity gauge, and thermometer to manage drying.

- Use supplemental heat only when needed to continue drying during high humidity conditions.

- Install adequately sized fans and air distribution ducts.

Bin drying can also be used to finish the drying process after rice has been dried to 16 to 18 percent moisture content in a column dryer. About 80 percent of rice in California is dried with column dryers. Details of two-stage drying are described in the warehouse drying section of this chapter.

The amount of time required for drying in bins is controlled by the airflow rate through the rice. The main goal of bin drying is to move as much air through the rice as possible. This is done by providing adequate airflow capacity, cleaning the rice, managing rice depth, and using supplemental heat during periods of prolonged high humidity.

Dryer Operation

Successful operation of bin dryers requires a calibrated rice moisture meter, a humidity sensor and a means to calibrate it, and a calibrated thermometer. These instruments guide the operator in making decisions on how deep to fill the rice in a bin and whether to use supplemental heat. More details on moisture meters are available in chapter 7, "Sampling and Measuring Rough Rice Moisture Content"; additional information on temperature and humidity measurement can be found in chapter 2, "Air Temperature, Humidity, and Weather."

Clean before drying

The first step in bin drying is to clean the rice with a high-efficiency aspirator and scalper before loading rice into the bin. Fines reduce the airflow through rice, causing increased drying time, and they may also cause wet spots if they are concentrated in a small volume of rice. Fines are often concentrated under the fill spout in the center of the bin. If present, they can be removed by unloading the center cone of rice from the bin, passing the rice through a scalper and aspirator and returning it to the bin. Grain spreaders reduce the buildup of fines in the center of a bin but pack the rice and reduce airflow. Operating stirrers can decrease the density of the rice after filling a bin with a spreader.

Table 6.1. Minimum airflow rate for bin drying of rice in California

Rice moisture entering bin (%)	Minimum airflow (cfm/cwt)
26	4.0
24	3.5
22	3.0
20	2.5
18 or lower	2.0

Source: Henderson and Parsons 1975.

Table 6.2. Typical fill depths for bin drying rice in California

Moisture content % wet basis	Depth of rice (ft)
< 18	up to 20
18–20	10
20–22	8
22–24	6
24–26	5
26–28	4
28–30	3

Source: Henderson and Parsons 1975.

Fill depth

The moisture content of rice limits the filling depth of a bin. High-moisture rice requires high airflow rates per hundredweight (cwt) to dry it before damage occurs (table 6.1). For example, rice at 26 percent moisture needs at least 4 cubic feet per minute per hundredweight (cfm/cwt). Bin drying operations rarely have the ability to change fan capacity, so airflow rate per hundredweight is controlled by adjusting the fill depth of the rice. As rice depth decreases, airflow from the fan increases slightly and less rice is dried, resulting in greater airflow per hundredweight. Table 6.2 shows the maximum rice fill depth for typical bin drying operations. Medium grain rice is harvested at a moisture content ranging from 18 to 24 percent, so bins are typically filled 5 to 10 feet deep.

Calculating the Amount of Rice in a Storage Bin or Warehouse

The first step is to calculate the volume of the rice using the geometrical formulas below. Remember to subtract the volume of air ducts in a flat storage. Next, multiply the rice volume times the test weight of the rice times a compaction factor. California medium grain rice has a test weight in the range of 36 to 38 pounds per cubic foot. Most moisture meters measure test weight along with moisture content. The best accuracy is obtained when the test weight is measured at storage moisture content. The compaction factor accounts for the packing caused by the falling grain as it is loaded into the storage and the settling that naturally occurs when rice is stored in a deep pile. The compaction factor increases as rice depth increases. It equals 1.14 for a flat storage and about 1.10 for a small metal grain bin.

Rice weight (lb) = volume (ft³) × test weight (lb/ft³) × compaction factor

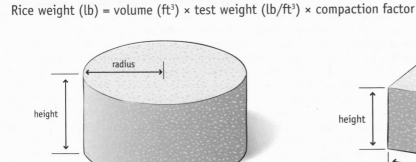

height × 3.14 × radius × radius = volume (ft³)
r = radius = distance around bin ÷ 6.28

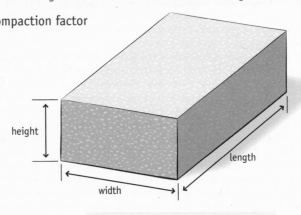

length × width × height = volume (ft³)

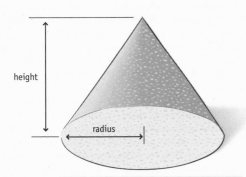

height × 1.05 × radius × radius = volume (ft³)

r = radius = diameter ÷ 2
OR
r = radius = distance around bin ÷ 6.28

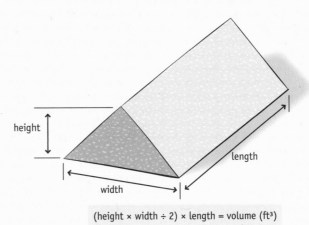

(height × width ÷ 2) × length = volume (ft³)

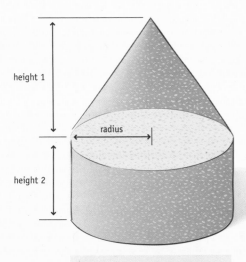

Calculate cone volume with height 1
Calculate cylinder volume with height 2

Add the 2 volumes to get total volume

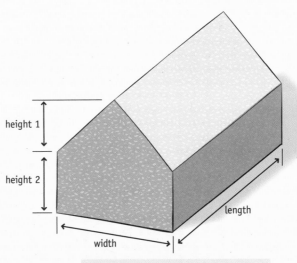

Calculate prism volume with height 1
Calculate rectangular volume with height 2

Add the 2 volumes to get total volume

Fan operation

Fans should be operated continuously until the rice moisture content drops below about 16 percent. When rice dries below 16 percent moisture content, operate fans when the outside relative humidity is below a threshold value that depends on the air temperature (see table 6.3). This approach is best implemented using an automated fan controller. For manual control, the approach can be approximated by operating the fan when relative humidity drops below 75 percent. If rice will be used immediately it should be dried to 13 to 14 percent moisture content and then transferred to a final storage bin.

If drying occurs late in the drying season and rice will be stored over the winter, rice can be transferred at 16 percent to the final storage bin. Running the rice through an aspirator during the transfer removes fines and prevents them from causing poor airflow in final storage. (See chapter 8, "Storage and Aeration," for details on holding rice at an elevated moisture content.)

Dryers with stirrers usually have no defined drying front, and rice moisture content can be measured with a sample taken at a random location in the bin, assuming the rice was well mixed by the stirrers just before the sample was collected. Bins without stirrers have a defined drying front, and the average moisture must be determined with a sample comprised of rice collected at regular intervals throughout the depth of the bin.

The amount of time for a drying front to move through a batch of rice increases with

- lower final moisture content
- lower airflow rate
- higher initial rice moisture content
- higher rice depth

Based on the airflow rate recommendations in table 6.1, a drying cycle requires from 1.5 to 4 days, assuming average September temperature and relative humidity conditions. At the end of the harvest season, a drying cycle may take three times longer because of cooler temperatures and higher relative humidity. Experience is the best way to estimate drying time for rice dried in a specific bin and fan system.

If stirrers are not used, drying time can be estimated by tracking the movement of the drying front through the bin. As the drying front moves through the batch, undried rice above the drying front will be several degrees cooler than the dried rice below the drying front.

Stirrer operation

If stirrers are used, fill the bin to a depth of at least 2 to 3 feet so the stirrers can operate without whipping. During drying, some operators prefer to continuously operate stirrers. Other operators reduce electricity use and equipment wear by operating them once to several times per week for long enough to completely mix the rice. Do not put high-moisture rice on top of dry rice with stirrers in operation. Mixing undried rice from the field with rice below 18 percent may cause reduced head rice quality.

Supplemental heat

Periods of continuous high relative humidity slow drying. During a rainstorm or when humidity is high for many hours per day because of a high ambient dew point temperature, air can be heated with direct-fired gas burners to continue drying. However, heat is not a substitute for adequate airflow; its main value is in dropping the relative humidity of the drying air, allowing fans to be operated even when the relative humidity of outside air is high.

Gas burners should raise drying air temperature by no more than 10° to 15°F. This may require adjusting the gas pressure regulator, changing the burner orifice, or replacing the burner if it cannot efficiently operate at a low heat output. If drying capacity is limiting harvesting operations, drying air temperatures can be increased by more than 10° to 15°F if outside air temperature is less than 70°F, but not beyond a maximum temperature of 85°F. Watch for condensation forming on the underside of the roof, roof supports, or the inside of exterior walls. If this occurs, reduce the air temperature rise until condensation stops.

Figure 6.1. Components of a bin drying system.

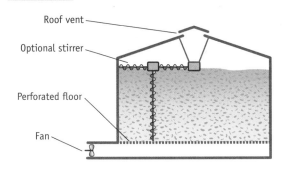

Roof vent

Optional stirrer

Perforated floor

Fan

When heat is used, measure the air temperature rise through the fan and burner. Measure the air temperature away from the burner flame so the thermometer is not influenced by the radiant heat from the flame. To measure the heated air temperature, drill a hole in the duct between the burner and the bin. Fit the hole with a plug so it can be sealed after taking temperature measurements.

Increasing air temperature beyond recommended levels results in overdrying the bottom layers of rice in a bin without stirrers. For example, adding 20°F to air at 100 percent relative humidity at typical ambient temperatures reduces relative humidity to 50 percent and causes the potential for overdrying the rice. Adding even more heat to increase the temperature rise only decreases the relative humidity, further increasing the potential for unnecessary over-drying. Adding excessive heat also increases the potential for condensation on the underside of the roof, increases energy costs, and can reduce head rice yield.

Bin Drying Equipment

Floors and ducts

In bin drying, fans supply air to the open space below a perforated floor that in turn distributes air to the rice (fig. 6.1). Bin floors should have an open area equal to about 15 percent of the total floor area. Air ducts above a solid floor should not be used because this system does not produce uniform airflow through the rice and is difficult to use when the bin is filled and emptied many times during the drying season.

Fans

Each drying bin should have a fan (in some cases more than one fan can be used) with adequate capacity, at least 4 cubic feet per minute per hundredweight to handle exceptionally high moisture content rice (table 6.1). Fan capacity is based on the amount of rice placed in the bin during drying; the airflow needed for storage aeration is much lower than that needed for drying. Both centrifugal (sometimes called squirrel cage) fans and axial flow fans are used. Centrifugal fans are quieter and usually operate against higher static pressures than axial flow (fig. 6.2). Backward incline centrifugal fans should be used when static pressures exceed 4.5 inches of water column (in. w.c.). Axial fans deliver more

Figure 6.2. Air output (cfm) as a function of static pressure (in. w.c.) for typical 5-horsepower fans.

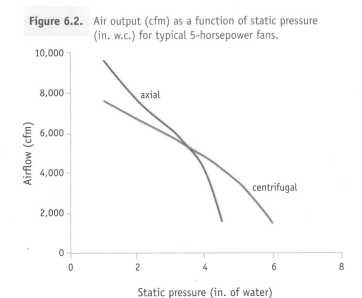

Axial flow fan used for bin drying and aeration. *Photo:* Robert Parsons.

Table 6.3. Outside air conditions for fan operation during drying and storage

Outside air temperature (°F)	Rice moisture content (%)	Outside air relative humidity (%)		
		To dry: Turn fan on only below	To aerate: Turn fan on between*	To add moisture: Turn fan on between
40°–49	16–17	78		
	15–16	73		
	14–15	67		
	13–14	57	57–66	66–92
	12–13			55–85
	11–12			50–78
	10–11			38–71
	9–10			27–64
50–59	16–17	82		
	15–16	76		
	14–15	70		
	13–14	61	61–69	69–92
	12–13			63–85
	11–12			53–78
	10–11			43–71
	9–10			33–64
60–69	16–17	83		
	15–16	78		
	14–15	72		
	13–14	64	64–72	72–92
	12–13			64–85
	11–12			56–78
	10–11			48–71
	9–10			37–64
70–79	16–17	84		
	15–16	80		
	14–15	73		
	13–14	67	67–73	73–92
	12–13			67–85
	11–12			60–78
	10–11			50–71
	9–10			40–64
80–89	16–17	84	do not	
	15–16	81	aerate at	
	14–15	75	temperature	
	13–14	69	over 79°	75–92
	12–13			68–85
	11–12			64–78
	10–11			55–71
	9–10			44–64
90–99	16–17	86	do not	
	15–16	82	aerate at	
	14–15	76	high air	
	13–14	72	temperature	77–92
	12–13			71–85
	11–12			66–78
	10–11			58–71
	9–10			45–64
100–109	16–17	88	do not	
	15–16	83	aerate at	
	14–15	79	high air	
	13–14	74	temperature	79–92
	12–13			73–85
	11–12			69–78
	10–11			61–71
	9–10			46–64

Source: Adapted from unpublished data by George E. Miller Jr. *Note:* *This range can be safely expanded by 3% below and above if needed for short-term aeration. If hot spots are a problem, fans and stirrers should be operated to cool rice and break up hot spots.

air than centrifugal fans at air pressures below 3.5 inches of water column. Either type of fan can be used at intermediate pressures.

Fans should be selected on the basis of their air volume at a specific static pressure. Use table 6.4 to estimate the pressure at the desired airflow rate. For example, a 21-foot-diameter bin filled to a 10-foot rice depth holds 1,410 hundredweight (table 6.5) and would require

about 5,640 cubic feet per minute assuming 4 cfm/cwt. Table 6.4 indicates that this fan would operate against a pressure of 2.5 inches of water column. Fan capacity and pressure can be used to select a fan from a supplier.

Increasing fan capacity decreases drying time and allows a bin to dry more rice per season. As a rule of thumb, doubling airflow decreases drying time by 50 percent, but it uses about 5 times more electricity per hour.

Fans vary greatly in their ability to convert electricity into airflow. The most accurate comparison is based on airflow in cubic feet per minute per watt of electricity consumption (cfm/watt). Efficient fans produce twice as much airflow per watt compared with inefficient fans. If cubic feet per minute per watt data is not available, the next best efficiency index is cubic feet per minute per horsepower (cfm/hp) based on the fan manufacturer's performance information. Motor nameplate power rating is not a reliable indicator of motor energy use because it is not usually operated at its rated capacity.

Table 6.4. Static pressure (inches w.c.) required for airflow through medium grain rice

Rice depth (ft)	Airflow (cfm/cwt)			
	2	3	4	5
20	5.0	8.2	12.0	16.0
18	3.9	6.8	9.4	12.6
16	2.8	5.2	7.0	9.9
14	2.2	3.7	5.5	7.3
12	1.6	2.7	3.8	5.0
10	1.2	1.6	2.5	3.6
8	0.6	1.1	1.4	2.0
6	0.4	0.6	0.8	1.0

Table 6.5. Capacities in hundredweight for round bins based on rice density of 37 lb/ft^3

Rice depth (ft)	Bin diameter (ft)											
	15	18	21	24	27	30	33	36	39	42	48	60
1	72	104	141	184	233	288	348	414	486	564	736	1,151
2	144	207	282	368	466	575	696	829	972	1,128	1,473	2,301
3	216	311	423	552	699	863	1,044	1,243	1,459	1,692	2,209	3,452
4	288	414	564	736	932	1,151	1,392	1,657	1,945	2,255	2,946	4,603
5	360	518	705	921	1,165	1,438	1,740	2,071	2,431	2,819	3,682	5,754
6	432	621	846	1,105	1,398	1,726	2,089	2,486	2,917	3,383	4,419	6,904
7	503	725	987	1,289	1,631	2,014	2,437	2,900	3,403	3,947	5,155	8,055
8	575	829	1,128	1,473	1,864	2,301	2,785	3,314	3,889	4,511	5,892	9,206
9	647	932	1,269	1,657	2,097	2,589	3,133	3,728	4,376	5,075	6,628	10,357
10	719	1,036	1,410	1,841	2,330	2,877	3,481	4,143	4,862	5,639	7,365	11,507
11	791	1,139	1,551	2,025	2,563	3,165	3,829	4,557	5,348	6,202	8,101	12,658
12	863	1,243	1,692	2,209	2,796	3,452	4,177	4,971	5,834	6,766	8,838	13,809
13	935	1,346	1,833	2,394	3,029	3,740	4,525	5,385	6,320	7,330	9,574	14,960
14	1,007	1,450	1,974	2,578	3,262	4,028	4,873	5,800	6,807	7,894	10,311	16,110
15	1,079	1,553	2,114	2,762	3,495	4,315	5,221	6,214	7,293	8,458	11,047	17,261
16	1,151	1,657	2,255	2,946	3,728	4,603	5,570	6,628	7,779	9,022	11,783	18,412
17	1,223	1,761	2,396	3,130	3,961	4,891	5,918	7,042	8,265	9,586	12,520	19,562
18	1,295	1,864	2,537	3,314	4,194	5,178	6,266	7,457	8,751	10,149	13,256	20,713
19	1,366	1,968	2,678	3,498	4,427	5,466	6,614	7,871	9,237	10,713	13,993	21,864
20	1,438	2,071	2,819	3,682	4,660	5,754	6,962	8,285	9,724	11,277	14,729	23,015

Fan blades with an airfoil cross-section use less energy than flat blades (fig. 6.3). The most efficient blades are usually made of cast metal rather than formed sheet steel. Efficient designs have a small distance between the tip of the blade and the fan housing. This minimizes the amount of air that recirculates at the blade tips. Recirculated air requires energy to move but does not contribute to airflow leaving the fan. Slow-turning, quiet fans use less energy than high-speed, loud fans. This generally means a larger-diameter fan operated at a low speed is more efficient than a smaller fan operated at high speed. Large-diameter fans can move air with less than half as much energy than smaller fans. Good air inlet and outlet design greatly increase fan efficiency. A sharp-edged inlet reduces fan output by 20 percent compared with a curved, aerodynamically shaped inlet (see fig. 6.3).

Stirrers

Stirrers mix rice during drying, keep it from packing, and allow about 10 percent more airflow than in rice that is not stirred. They can also be used to level rice after filling the bin, and during storage they can break up wet spots. However, stirrers have disadvantages, and some operators choose not to use them. They add capital and operating cost and reduce the holding capacity of a bin. Some operators prefer to invest in additional storage volume rather than buy stirrers. Experience in the Midwest indicates that stirrers have the greatest value when supplemental heat is added to the drying air, because they prevent overdrying of rice.

Drying and Cooling Fronts

In a bin without stirrers, rice dries in a layer called a drying front that moves slowly from the floor to the top of the batch (fig. 6.4). As the drying air moves up through the rice it gains moisture from the wet rice and cools. This causes the relative humidity of drying air to increase. When the air reaches the equilibrium relative humidity of the rice (the specific humidity depends on rice moisture and rice temperature), drying stops. Rice below the drying front is in

Figure 6.3. Characteristics of an efficient fan design.

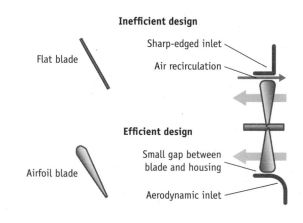

Figure 6.4. Zones of rice temperature and moisture formed during ambient air drying.

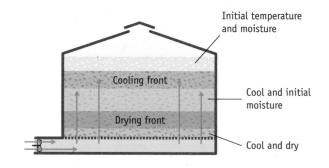

temperature and moisture equilibrium with the drying air.

Immediately above the drying front is a cooling front where the rice temperature drops but the rice moisture content remains near its initial level. The cooling front moves much faster through the rice than the drying front. Rice above the cooling front is at the temperature and moisture content of the rice when it was loaded into the bin. The cooling front does not exist for long because it moves so quickly through the rice. Most of the time, rice above the drying front has cooled but has not dried.

Factors Controlling Drying Rate in Bin and Warehouse Drying

Rice dries faster if it is exposed to

- higher airflow rates
- lower relative humidity
- higher air temperature

Airflow for dryers is best described in terms of cubic feet per minute per hundredweight of product (cfm/cwt). At low airflow rates, small increases in airflow greatly decrease drying time, but at high airflow rates, increases in airflow cause small decreases in drying time (fig. 6.5). Airflow rates over 50 cubic feet per minute per hundredweight are commonly used in column dryers where air travels through only 6 to 12 inches of rice. In bins, high airflow rates are not feasible because of the great amounts of fan power required to push high airflow rates through a deep bed of rice. Airflow in a filled bin of rice is often in the range of 1 to 2 cubic feet per minute per hundredweight. A 3- to 6-foot layer in a bin may allow airflow rates of about 5 cubic feet per minute per hundredweight.

In bin dryers or flat storage, maximum seasonal drying capacity is obtained by starting drying as early in the season as possible to take advantage of higher air temperatures and lower relative humidity. During September in the

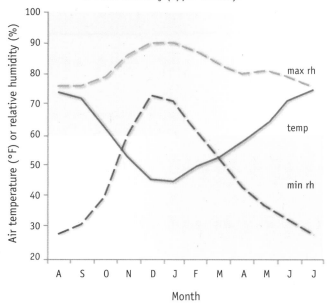

Figure 6.6. Monthly average conditions for Sacramento, California, air temperature (solid), minimum relative humidity (lower dashed), and maximum relative humidity (upper dashed).

rice-growing region of California, average air temperature ranges from 72° to 76°F (fig. 6.6). At the end of the drying season in November, monthly average air temperature drops to less than 55°F. Bin drying in November takes about three times longer than it does in September. By the end of November, the minimum relative humidity is greater than 60 percent, and in most years it is difficult to finish drying until the humidity drops again in the spring.

Gas burners can be used to heat the air and reduce relative humidity. The main value of heating is to allow fans to operate during periods of high relative humidity. This may allow drying to finish earlier in the fall, but adding heat when the outside air is already suitable for drying just tends to overdry the bottom layers of rice in bins without stirrers. Stirrers prevent the formation of an overdried layer and allow better use of heated air.

Air humidity is sometimes controlled separately from air temperature by using special dehumidification equipment. Some of these units supply air at a specified humidity and temperature and are useful for providing

Figure 6.5. Effect of airflow rate on drying time for rice dried from 24% to 18% moisture content with air temperature equal to September average for Red Bluff, California.

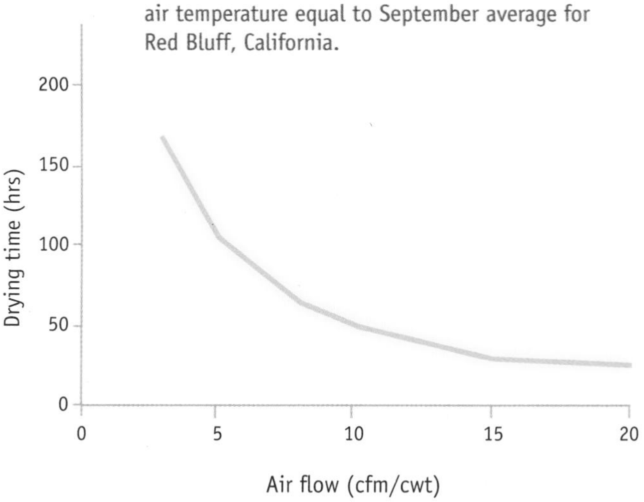

very controlled conditions for rice in storage. However, they are expensive to purchase and operate.

Increasing Airflow through Rice

Airflow (measured in cfm/cwt) can be increased by:

- decreasing rice depth
- installing a second fan
- using stirrers

Decreasing rice depth increases the airflow per hundredweight because the volume of rice is reduced, and fans produce slightly more air when operating against lower airflow resistance (fig. 6.7).

Shallow rice depths have one disadvantage: a certain amount of drying air is wasted. After the drying front reaches the top of the rice, the drying air does not gain its maximum amount of moisture before it leaves the rice; this air could have been used for additional drying if it were passed through more rice. Maximum seasonal drying capacity is achieved when there is always a layer of undried rice on the top of the bin. Some dryer operators achieve this by filling bins fairly deep, at least 10 feet, and drying until the topmost rice just begins to lose moisture. Then they empty the topmost layer of rice into another bin and add more undried rice to the bin. The

lower part of the first bin is transferred to a bin with dried rice.

It is possible to increase airflow by adding an additional fan. For a well-designed system, a second fan, which doubles fan horsepower, can increase airflow by 25 to 35 percent. The additional fan is usually placed in parallel with the existing fan so that both fans feed directly into the bin. Carefully consult the manufacturer's performance data to determine the combined performance of the fans. The fans should have nearly the same performance characteristics, otherwise it is possible for one fan to push air out the other.

Warehouse Drying

Nearly all rice dried in facilities with column dryers is put into warehouse storage at 16 to 18 percent moisture content; final drying is accomplished by moving unheated outside air through the rice. This type of drying takes many weeks to complete. It is also complicated by the nonuniform airflow common in warehouses with air supplied through tunnel ducts placed on the concrete floor.

Operation

Managers often fill the warehouse in layers of rice about 5 to 10 feet deep. While there may be logistical reasons for this, the shallow layers

Figure 6.7. Example of the effect of rice depth on airflow rate through rough rice in a grain bin, based on a typical centrifugal fan.

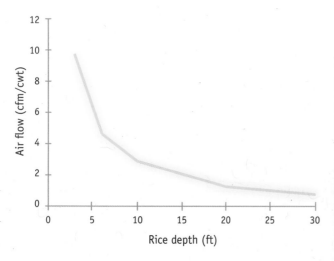

Airflow tunnels and temperature cables in warehouse storage. *Photo:* Randall Mutters.

do not speed drying. A shallow layer allows much of the drying air to travel from the tunnel directly upward to the top of the rice, missing the rice between the tunnels. The best airflow uniformity is maintained by completely filling the warehouse to the height of the eaves, usually 20 to 24 feet high. If the warehouse is not going to be completely filled, leave one end empty and fill the rest to the height of the eaves.

Storing the rice above eave height in a large row is sometimes done to obtain extra storage volume. But this causes poor airflow distribution in the rice, since little air flows to the rice near the top of the pile. The top rice is very difficult to dry and should be at safe storage moisture before entering the warehouse. Remove the peaked rice as soon as possible because of the difficulty of getting enough airflow for proper aeration during storage.

Fans are operated in the same fashion as they are for bin dryers. When the majority of the rice has dried below desired moisture content, the fans are operated when humidity conditions are suitable for aeration (table 6.5).

A periodic survey of rice moisture should be done to determine when the rice is dry. The survey must include samples across the entire warehouse and at three to four depths. Be sure to sample areas that are likely to be the driest (rice directly above the air tunnels) and those likely to be the wettest (rice between the tunnels and close to the floor). A complete survey may require collecting several hundred samples. The samples near the bottom of the warehouse require a vacuum probe to collect. Unfortunately, this is a time-consuming process, but it is necessary because large storage areas may contain rice from many different lots, and the uneven airflow adds to the potential for moisture content variability.

Temperature cables are a valuable tool in estimating the location of the drying front. Rice above the front is cool after the cooling front has moved through the batch. Rice below the drying zone is slightly warmer. Note that the drying front is not level across the rice: it is higher directly above the air ducts and lower between the ducts.

Equipment

Airflow in warehouse storage is provided by fans mounted near the floor. These fans must be sized to provide a minimum of 0.4 cubic feet per minute per hundredweight when the warehouse is filled. Some facilities have three to four times the minimum airflow to speed drying and aeration. Fans direct air to screen-covered air ducts (sometimes called tunnels) that run across the building (fig. 6.8). The cross-sectional area of the ducts should be large enough to limit air speed in the ducts to less than 2,000 feet per minute. Calculate duct size using the following formula:

$$\text{duct cross sectional area (ft}^2\text{)} = \text{airflow (cfm) / air speed (fpm)}$$

Figure 6.8. (A) Top view of aeration tunnels and fans in a warehouse; (B) cross-section of a warehouse showing airflow.

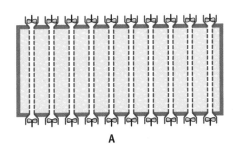

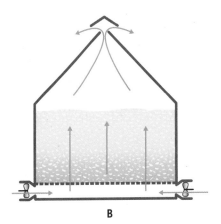

Unlike bins with perforated floors, the airflow distribution is very uneven across a warehouse storage. The airflow is greatest directly above the ducts and lowest near the floor, between the ducts. The low airflow between ducts causes slow drying in these areas (fig. 6.9), and the uneven drying is especially evident when the storage is not filled to the height of the eaves.

Figure 6.9. Schematic of rice in a warehouse showing areas of low airflow between air supply ducts.

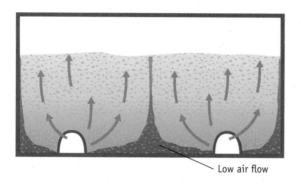

Low air flow

Figure 6.10. Recommended duct spacing in a flat warehouse.

Recommendation: L < 1.5 × H
and duct-to-duct spacing = H, duct to wall = H/2

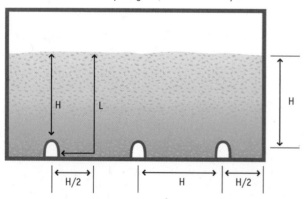

Ducts should be spaced so that the longest distance the air must flow is less than 1.5 times the shortest distance to minimize airflow nonuniformity (fig. 6.10). This means that the spacing between ducts should be equal to the rice height, and the distance between the outside ducts and the walls should be half the rice height.

Fans can be installed to force the air into the warehouse (a pressure system) or exhaust the air out of the facility (a suction system). The suction system prevents moisture condensation on the underside of the roof, but the wettest rice is near the floor and difficult to monitor. The pressure system is recommended because it allows easier management of the wettest rice. Some facilities are designed so air direction can be changed. Suction is used to reduce dust when people are filling the warehouse with rice, then after filling, the airflow direction is switched for drying.

References

Henderson, S. M. L., and R. A. Parsons. 1975. Deep bed grain drying. University of California Division of Agricultural Sciences Leaflet 2103.

Jayas, D. S., N. D. G. White, and W. E. Muir, eds. 1995. Stored-grain ecosystems. New York: Marcel Dekker.

Loewer, O. J., T. C. Bridges, and R. A. Bucklin. 1994. On-farm drying and storage systems. St. Joseph, MI: ASAE.

Midwest Plan Service. 1980. Low temperature and solar grain drying handbook. MWPS-22. Ames: Iowa State University.

Mossman, A. P. 1986. A review of basic concepts in rice-drying research. Critical Reviews in Food Science and Nutrition 25(1): 49–70.

Sauer, D. B., ed. 1992. Storage of cereal grains and their products. 4th ed. St. Paul: American Association of Cereal Chemists.

Sampling and Measuring Rough Rice Moisture Content

Rice moisture content is an essential factor in managing harvest, drying, and storage, and it determines the value of rice when it is sold. In marketing, the buyer and seller need an accurate measure of the average moisture content of the rough rice to determine the value of the rice. Water is very valuable in a grain. For example, a lot of 1,000 tons with 14 percent moisture has 140 tons of water, and at 13 percent moisture content, the lot contains 129 tons of water; at $0.10 per pound of rice the 11-ton difference is worth $2,200. Assuming a yield of 80 hundredweight per acre, the lost revenue equals $9 per acre when rice with the lower moisture content is sold. Both the buyer and seller need to know how much water they are trading. In harvesting and storage, the average moisture content of a lot is often used to describe handling procedures; however, high-moisture-content areas in part of the lot is just as important. For example, in storage, fungal damage can occur in rice with the highest moisture content.

Estimating Rice Moisture Content

Good estimates of rice moisture require the following steps.

- Collect a representative sample:
    * collect grain from whole plants in the field
    * use a grain probe in trucks
    * collect multiple samples from the column dryer
    * use a vacuum probe in a warehouse

- Use a splitter to obtain a representative sample that will fill the moisture meter.

- Use a properly calibrated electronic moisture meter.

- Correct for meter error when using a capacitance type (GAC 2100) meter with rice just collected from a column dryer.

Both proper sample collection and accurate moisture content determination are required for useful moisture content measurement and preserving the value of the rice.

Sample Collection

At Harvest

Determining when to harvest requires estimating the average moisture content and the moisture content variation in the field. Unfortunately, there are often large differences in rice moisture content across a field (fig. 7.1). A composite sample collected from the combine is the fastest way to get a good sample for estimating average moisture content of a field. However, growers do not want to incur the expense of bringing in a combine if a field is not close to harvest moisture content.

Hand-collected samples should be obtained from locations with a range of moisture contents. The driest rice will be in areas that mature first and can sometimes be detected by lighter grain color. Wet areas mature last and may be located in borrow areas near the levees and areas that drain last. Pick complete panicles and strip all of the kernels. Collecting all panicles

Every incoming load is sampled for moisture content. *Photo:* James Thompson.

from a plant is a more accurate but more time-consuming method; it ensures that the sample includes kernels from main and secondary tillers in their actual proportion. Hand-sampling usually underestimates moisture content by several percentage points.

Receiving at the Dryer

Drying costs are based on average moisture content, but it varies a great deal in a truckload of rice, and a sample must be collected throughout the load to get a good estimate of the average moisture content. For example, the field depicted in figure 7.1 has moisture contents ranging from 10 to 22 percent. It is possible for a load of rice to have sections that vary in moisture content by well over 10 percentage points. Collecting a single sample from the grain as it is being dumped is a poor way to sample. Using a grain probe reduces variability in sampled moisture content. For best results, use a grain probe and sample in several locations. The sample will be too large to put in a moisture meter, so it should be split with a commercial splitting device. Do not just grab a can full of rice out of a bucket of sampled rice. If a splitter is not available, mix the sample

Figure 7.1. Variability of moisture content in a field of rice. *Source:* Marchesi 2005.

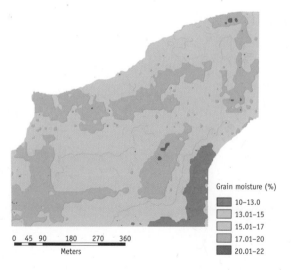

Grain moisture (%)

- 10–13.0
- 13.01–15
- 15.01–17
- 17.01–20
- 20.01–22

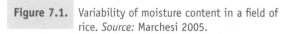

0   45   90   180   270   360
Meters

Figure 7.2. Distribution of single-kernel moisture contents for M-202 rice with 22.8% average moisture content.

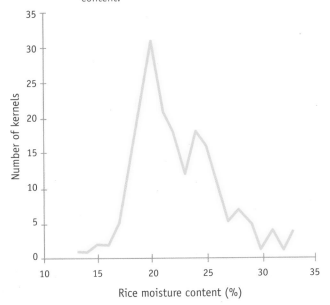

A splitter ensures a representative sample for moisture content measurement. *Photo:* Michael Poe.

thoroughly by hand or by repeatedly pouring between buckets before taking the moisture meter sample.

During Column Drying

Operators need to know the average moisture of rice going into and out of the dryer. Collect samples from the dryer supply and discharge conveyors every half hour to hour during an 8-hour lot and average the sample moisture contents. Remember that the commonly used capacitance-type moisture meters read several percentage points low when measuring rice flowing out of the dryer. This error can be eliminated by tempering the samples in sealed containers for several hours before determining their moisture content.

In Storage

Rice is often not completely dry when it is put in storage. In a large, flat storage warehouse, rice moisture content may vary considerably, particularly when the rice is only partially dried. Sample the entire warehouse every 2 to 4 weeks during dry-down. Drying is managed based on the wettest rice, and moisture content must be

determined on each sample collected. Drying continues until the wettest areas have reached a safe moisture content. Use a vacuum probe to sample at three to four depths from the top of the rice to the floor. Be sure to collect samples from along walls and between the air tunnels, where moisture is likely to be highest since they receive the least amount of airflow. Areas that have been steadily heating because of insect or mold growth are also likely locations of high moisture content. Measure the moisture content of each sample and plot the data, looking for wet spots and areas that have reached desired storage moisture content. Intensive sampling like this may also be needed as the rice warms later in the spring, especially if off odors or insects are detected in the storage. If insect activity is suspected, each sample should also be sieved. Even if the rice is of uniform moisture content

Sampling stored rice with a vacuum probe.
Photo: Randall Mutters.

after drying in the fall, later in the storage season differences in temperature within the grain mass, insect activity, and mold growth may result in areas of high moisture content.

Measuring Moisture Content

The official USDA-GIPSA moisture meter is the DICKEY-John GAC 2100 machine, superseding the Motomco 919. Properly calibrated machines correlate closely to air oven determination of moisture content if the samples have been stored for some time before analysis. However, in everyday use, the variability is greater. The moisture content variation in field tests of two meters was high enough that any individual reading may be 2 percentage points of moisture different than the air oven standard. The variation may have been caused by the following conditions:

- field samples having a wide range of individual kernel moisture content (see fig. 7.2)

- nonuniform moisture in individual kernels because they have just been subjected to drying or rehydration

- contamination with weed seeds or other foreign material

- moldy or diseased kernels

Good sample handling procedures before moisture measurement reduce the error caused by the factors mentioned above. Clean rough rice samples that have not been recently subject to moisture change or temperature change provide the least potential for measurement error. Plastic bags are generally not moisture tight and should not be used to hold samples for more than 4 hours. Protect samples from direct sunlight and other sources of heat. Samples with condensation on the inside of the sample container should not be placed in a meter; reseal the containers and store them until the condensation disappears. Samples collected in the field and transported for measurement can be protected from heat by placing them in an insulated food cooler.

Samples collected directly from a dryer are not at moisture equilibrium, and if they are placed in a GAC 2100 meter immediately after collection they may produce a moisture reading several percentage points lower than the actual moisture. Dryer operators call this effect "bounce" because the moisture content appears to bounce up after subsequent storage. The error can be minimized by holding the sample in a sealed container for at least 4 hours before placing it in a meter. If that is not possible the error can be estimated with a chart (table 7.1). Single-kernel moisture meters do not have this error and can measure samples directly from the dryer.

The methods of determining moisture content of rice grain can be divided into two broad categories: direct and indirect (fig. 7.3). Direct methods determine the water content by removing the moisture. For example, oven methods evaporate the moisture from the grain and determine the water content by the weight loss. Indirect methods, in contrast, measure an electrical property of the grain or equilibrium relative humidity. The direct methods provide a standard measurement of moisture content and are used to calibrate the more practical and faster indirect methods.

Direct Methods

Chemical reaction

The most accurate method consists of extracting the water chemically from the rice. The chemical method uses the reaction of iodine with water in the presence of sulfur dioxide. It is seldom used because it is time-consuming and costly.

Oven heating

Several organizations publish an air oven standard for testing the moisture content of ground rice: heating a 2-gram sample for 1 hour at 266°F. The moisture content of the grain is determined using the formula below. The ISO 712 air oven standard for whole kernels is 266°F for 24 hours. Heating whole kernels at 221°F for 72 hours also produces consistent moisture content determinations. Vacuum oven methods are also available.

$$M_w = [(W_{initial} - W_{final}) \div W_{initial}] \times 100$$
where:

M_w = moisture content (%) on a wet weight basis (the standard basis used by the industry)

$W_{initial}$ = initial weight of sample before drying (grams or pounds)

W_{final} = final weight of sample after drying (grams or pounds)

Distillation

Ground grain samples can be mixed with toluene and boiled. The water comes off as a vapor; it is condensed and the volume measured, allowing the moisture content of the grain to be determined.

Infrared and microwave radiation

Both infrared and microwave radiation are used to evaporate the water in a grain sample. Determining the moisture content based on either type requires the use of ground samples. Exposure to the radiation must be carefully controlled to prevent the sample from burning.

Indirect Methods

Electrical resistance

Electrical resistance moisture meters measure a grain sample's resistance to an electrical current. The level of the resistance is correlated to moisture content. A whole grain sample is placed between two electrodes in a compression cell. The sample must be compressed to a known and

Figure 7.3. Classification of grain moisture content (MC) measurement methods.

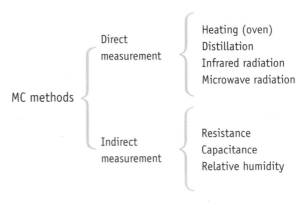

Table 7.1. Moisture correction for GAC 2100 used to measure rough rice directly from a column dryer; correction is based on California medium grain rice dried with a discharge temperature of 90° to 95°F

GAC 2100 reading (%)	Actual moisture* (%)	Add to GAC moisture (%)
13.0	12.5	−0.5
13.5	13.4	−0.1
14.0	14.2	0.2
14.5	15.1	0.6
15.0	16.0	1.0
15.5	16.8	1.3
16.0	17.7	1.7
16.5	18.6	2.1
17.0	19.4	2.4
17.5	20.3	2.8
18.0	21.2	3.2
18.5	22.0	3.5

Note: *Measured after sample had been stored in a sealed container for more than 4 hours.

constant value for accurate measurements; a grain temperature correction is sometimes used. A few hand-held grain meters use electrical resistance. Single-kernel moisture meters also use resistance measurements to determine moisture content.

Electrical capacitance

With electrical capacitance meters, a sample is poured into an enclosure with metal walls and a center electrode that form the conductors of a condenser activated by a high-frequency current. Accurate measurements require correction for grain temperature and bulk density. A separate calibration is needed for every grain type. Capacitance meters are generally more accurate over a wider range of moisture contents than are resistance meters. The Motomco 919 and the Dickey-John GAC 2100 meters are examples of this type of meter.

Relative humidity

Other meters measure the equilibrium relative humidity (ERH) in the air space between the grains. The ERH in the air surrounding a grain sample depends on the moisture content of the grain. Various types of hygrometers are used to measure the relative humidity. Correction for temperature is required. Good accuracy from this measurement requires that the sample be well tempered. Each grain type has a unique relationship between equilibrium humidity and moisture content.

Shrink Factors

Rice is weighed when it is received at the dryer, but the weight of the lot will be affected by the drying process. Shrink charts or formulas predict the weight of rice that will be left after it has been cleaned and dried. Shrink calculations reduce the received weight by factors that account for moisture lost in drying, dockage removed in cleaning and handling operations, and invisible loss associated with grain respiration and loss of fine material from individual grains.

Weight loss associated with moisture loss is calculated using the following formula:

$$S = (1 - [(100 - M_i) \div (100 - M_f)]) \times 100$$

where:
S = shrink factor (%)
M_i = initial moisture content (%)
M_f = final moisture content (%)

When shrinking to 13 percent moisture content, the formula can be simplified to:

$$S = (M_i - 13.0\%) \times 1.15\%$$

For example, if rice at receiving has a moisture content (M_i) of 20 percent, it will lose $(20 - 13) \times 1.15 = 8.05$ percent of its original weight because of moisture loss in drying.

Weight loss associated with dockage is assumed to be a constant factor and is commercially set at a level of about 2.0 percent. Invisible loss values range from 1.5 percent to 3.5 percent and are also set by the receiving company. Dockage and invisible loss can be calculated based on the received weight or the weight after the rice has been dried. The first method assumes that dockage and invisible loss are weighed before drying. The second method of calculation assumes that their weight is determined after drying.

Tables 7.2 and 7.3 illustrate the effect of all the factors and assumptions on total shrink. Notice that the shrink factor is larger when the dockage is subtracted from the received weight before drying. In some charts, invisible loss is not subtracted from rice received at 13 percent or lower moisture. Check with your dryer to get the shrink chart that will be used for your rice.

Table 7.2. Sample shrink chart for rice; calculations assume a 13% final moisture content and 2% dockage and 2% invisible loss, each applied to the received weight

Receiving weight (lb)	Receiving moisture (%)	Moisture shrink (lb)	Dockage (lb)	Invisible loss (lb)	Total shrink (lb)	Final weight (lb)
100	13	0.00	2	2	4.00	96.00
100	14	1.15	2	2	5.15	94.85
100	15	2.30	2	2	6.30	93.70
100	16	3.45	2	2	7.45	92.55
100	17	4.60	2	2	8.60	91.40
100	18	5.75	2	2	9.75	90.25
100	19	6.90	2	2	10.90	89.10
100	20	8.05	2	2	12.05	87.95
100	21	9.20	2	2	13.20	86.80
100	22	10.35	2	2	14.35	85.65
100	23	11.50	2	2	15.50	84.50
100	24	12.65	2	2	16.65	83.35
100	25	13.80	2	2	17.80	82.20
100	26	14.95	2	2	18.95	81.05
100	27	16.10	2	2	20.10	79.90
100	28	17.25	2	2	21.25	78.75
100	29	18.40	2	2	22.40	77.60
100	30	19.55	2	2	23.55	76.45

Reducing Variability in Quality Appraisal Samples

Variability and error in appraisal samples can be minimized by

- collecting a representative sample; do not use a single catch can sample

- drying samples with room temperature air to maximize head rice quality

- drying samples to a constant moisture content, because low-moisture samples have slightly higher head rice quality than samples at 14 percent moisture content

An evaluation of variability at the California Department of Food and Agriculture Grain Inspection Laboratory in West Sacramento showed that head rice yield results have a standard deviation of 1.2 percentage points (fig. 7.4). This means that 68 percent of replicate sample results will be within a range of 2.4 percentage points and that 95 percent of the results will be within a range of 4.8 percentage points. Testing has also shown that results change slightly if samples were evaluated in the first few days after drying and did not change after longer storage periods.

Sample Collection

To get a representative sample of a lot, use a standard multisample vacuum probe and use a splitter to obtain the needed sample amount.

Sample Drying

Air temperature used for sample drying can affect head rice quality. Maximum quality is obtained by using air at a constant room temperature of 75°F or lower (see fig. 7.4). If the air is heated, the rice should be exposed to the warmed air only periodically and allowed to temper between exposures. For example, the California Warehouse Association recommends heated air at 100°F for 30 minutes followed by a 4-hour tempering before the next 30-minute exposure. This procedure produces about 2.5 percentage points lower loss in head rice yield than does drying the appraisal sample at room temperature.

Table 7.3. Sample shrink chart for rice; calculations assume 13% final moisture content and 2% dockage and 2% invisible loss, each applied to the weight of rice after it has been dried to 13% moisture

Receiving weight (lb)	Receiving moisture (%)	Moisture shrink (lb)	Dockage (lb)	Invisible loss (lb)	Total shrink (lb)	Final weight (lb)
100	13	0.00	2.00	2.00	4.00	96.00
100	14	1.15	1.98	1.98	5.10	94.90
100	15	2.30	1.95	1.95	6.21	93.79
100	16	3.45	1.93	1.93	7.31	92.69
100	17	4.60	1.91	1.91	8.42	91.58
100	18	5.75	1.89	1.89	9.52	90.48
100	19	6.90	1.86	1.86	10.62	89.38
100	20	8.05	1.84	1.84	11.73	88.27
100	21	9.20	1.82	1.82	12.83	87.17
100	22	10.35	1.79	1.79	13.94	86.06
100	23	11.50	1.77	1.77	15.04	84.96
100	24	12.65	1.75	1.75	16.14	83.86
100	25	13.80	1.72	1.72	17.25	82.75
100	26	14.95	1.70	1.70	18.35	81.65
100	27	16.10	1.68	1.68	19.46	80.54
100	28	17.25	1.66	1.66	20.56	79.44
100	29	18.40	1.63	1.63	21.66	78.34
100	30	19.55	1.61	1.61	22.77	77.23

Note: Totals do not sum due to rounding.

Sample Moisture

Sample moisture content at appraisal milling affects head rice yield. For example, medium grain rice gains about 2 percentage points of head rice yield when the sample moisture drops from 13.5 percent to 12 percent (see fig. 7.4). Short grain rice is less affected by sample moisture; long grain rice may have a 6-point difference over this moisture range.

Summary

Key management decisions and, in the end, the value of rice depend on representative and accurate determinations of moisture content.

Each organization should develop a clear set of procedures for measuring moisture content. The procedures should specify methods of sample collection and handling for each kind of sample collected, as well as the meters or moisture content measurement techniques to be used. Standard procedures produce repeatable results and allow managers to compare data from year to year and between various operations. But even with standard procedures, managers must understand that all commercially feasible moisture content measurements have error, and management decisions based on moisture content must account for a degree of uncertainty.

Figure 7.4. Variability and error in head rice yield results associated with appraisal sample collection, sample drying method, and sample analysis.

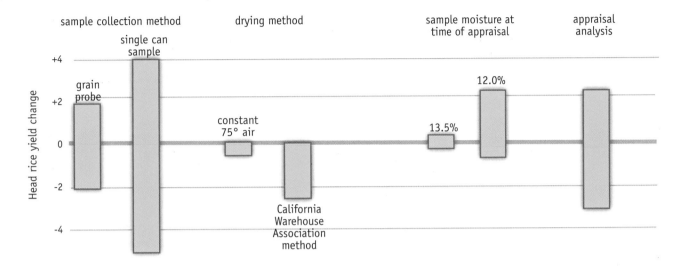

References

Brooker, D. B., F. W. Bakker-Arkema, and C. W. Hall. 1992. Drying and storage of grains and oilseeds. New York: Van Nostrand Reinhold.

Chen, C. 2003. Evaluation of air oven moisture content determination methods for rough rice. Biosystems Engineering 86(4): 447–457.

Marchesi, C. E. 2005. Association between rice grain quality, grain moisture, and soil characteristics in California leveled fields. MS Thesis. University of California, Davis

Sauer, D. B., ed. 1992. Storage of cereal grains and their products. 4th ed. St. Paul: American Association of Cereal Chemists.

Thompson, J. F., and R. G. Mutters. 2001. Maintaining rice quality after harvest. In Annual comprehensive rice research, University of California. Yuba City: California Rice Research Board. CARRB Web site, http://www.carrb.com/AnnualRpts.htm.

Thompson, J. F., J. Knutson, and B. Jenkins. 1990. Analysis of variability in rice milling appraisals. Applied Engineering in Agriculture 6(2): 194–198.

Michael Poe

Storage and Aeration

Introduction

A properly operated storage keeps rice at a prescribed moisture content and temperature in order to preserve its quality and keep it free of insects and decay. Rice moisture content and temperature is controlled by regularly forcing outside air through the stored rice. This process is called grain aeration. Aeration fans are operated when outside air temperature and relative humidity are at levels that will achieve the desired rice conditions. A key to understanding how air temperature and humidity control rice moisture content is the concept of equilibrium moisture content.

Equilibrium Moisture Content

Rice gains or loses moisture based on its moisture content and the temperature and humidity of the air around it. For example, if rice is placed in an airstream that has a constant temperature and constant relative humidity, the rice will attain an equilibrium moisture content corresponding to the relative humidity and temperature of the air (fig. 8.1). This is how rice is dried in drying operations. However, in storage, the dried rice is

Good Storage Practices to Prevent Rice Quality Loss

- Keep rice below a moisture content that prevents mold growth.

- Keep the temperature of rice below 60°F as long as possible to prevent insect activity.

- Store only well-cleaned rice.

- Store rice with a level top surface rather than a peak.

- Design and operate the aeration system to maintain uniform rice moisture content and temperature.

- Regularly inspect rice for mold, insects, warm rice, and wet spots.

aerated with outside air in order to cool it during the fall and early winter months or warm it in the late spring and summer. In the fall, cool air comes in contact with the warm grain, causing it to cool and air to warm. Before the air used for aeration leaves the stored rice, its temperature and relative humidity come into equilibrium with the rice; that is, the relative humidity and temperature of aeration air correspond to the equilibrium temperature and humidity of the rice.

Aeration Fronts

Aeration is the process of exposing stored rice to outside air in order to equalize the temperature of rice and reduce the temperature difference between the rice and the outside air. The aerating effect of air moves through a bin of rice in a front, similar to the way cooling moves through rice in a front (see chapter 6, "Bin and Warehouse Drying"). For example, before a bin of rice is aerated, the air and rice in the bin are in temperature and moisture content equilibrium. If the outside air used for aeration is cooler than the rice (and air) in the bin, the rice at the bottom of the bin will cool first, and then the rice immediately above it will cool, and so on up through the bin. This process can be viewed as an aeration front moving through the

Table 8.1. Approximate time for complete aeration with various airflow rates

Airflow rate (cfm/cwt)	Time for a complete aeration (hr)
0.2	120
0.5	50
1.0	25
1.5	15

rice. An aeration cycle is completed when the aeration front has moved all the way through the rice. The time required for an aeration front to move through rice ranges from 15 to 120 hours, depending on fan capacity (table 8.1.)

Safe Storage Moisture and Temperature

At temperatures near 70°F, the growth of fungi (mold) is minimal below 65 percent relative humidity, which is equal to a water activity of 0.65 (fig. 8.2) and a moisture content of 13.5 to 14 percent (for more information on water activity, see chapter 10, "Quality Changes during Handling and Storage"). The growth of yeast and bacteria is minimal at even higher relative humidities. As rice temperature drops below 70°F, the safe storage moisture increases. At 50°F, fungi do not grow below 75 percent relative humidity, corresponding to a rice moisture content of 15 to 16 percent. Commercial rice warehouses in California sometimes store rice at 16 percent moisture content over the winter months with little noticeable mold growth as long as the rice is kept uniformly cool and carefully monitored for signs of mold.

Keep the temperature of stored rice below 60°F to minimize insect damage (see fig. 8.2). Insects become active when rice temperature rises above 60°F, and infestation and damage are likely when rice reaches 70°F. If rice is held during the summer without a well-operated aeration system or refrigerated storage it will usually need fumigation to control insects.

Moisture Migration during Storage

Rice in storage is subject to moisture migration caused by differences in grain temperature. This is particularly true for grain stored in metal bins

Figure 8.1. Relationship between temperature and relative humidity of air and equilibrium moisture content of rough rice. *Source:* ASAE 1995.

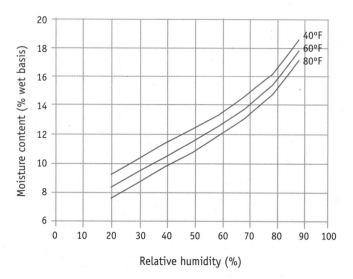

Figure 8.2. Critical threshold conditions for the growth of mites, insects, and mold.

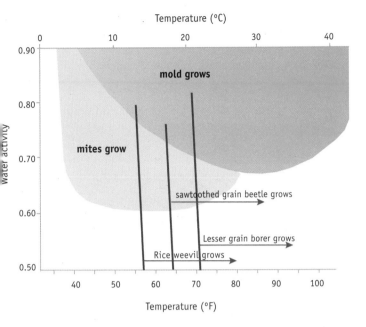

Figure 8.3. Typical locations of wet rice during cold ambient conditions.

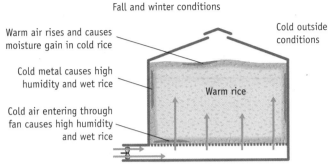

air temperature conditions for two Sacramento Valley locations. In July and August average air temperature exceeds 70°F, 10°F more than the maximum desirable storage temperature of 60°F. This is the most difficult period to store rice, and it must be monitored regularly to check for signs of insect activity.

Aeration System Operation

Optimal aeration control strategies keep rice at a prescribed, uniform temperature and moisture content during storage using a minimum of electricity and labor. Unfortunately the relationships between rice moisture and temperature, insect growth, decay development, and meteorological conditions are quite complicated. Optimal aeration control strategies have not been thoroughly researched for California stored rice. The recommended options presented below are guidelines that have worked in commercial practice, but further research may provide the industry with even better approaches.

compared with concrete-walled storages. Figure 8.3 shows the likely locations of wet spots. In the late fall and early winter, stored rice tends to be warmer than the outside air. Warm air rises slowly out of the center rice, and when this air contacts cold rice on the top of the bin the air cools, increasing its relative humidity and causing the top rice to gain moisture. Rice close to cold exterior walls and near aeration fans and ducts can also be cold in the winter and may gain moisture. When improperly aerated bins are emptied, they sometimes have moldy rice clinging to the inside of the walls. Improperly aerated flat storages sometimes have moldy rice near aeration ducts and corners.

Aeration minimizes moisture migration by equalizing the rice temperature in the storage and keeping the rice temperature close to the average conditions of the outside air. In the late fall, winter, and early spring, the rice temperature should be within 10°F of the average outside air temperature. Figure 8.4 shows the average

Figure 8.4. Monthly average air temperature in the Sacramento Valley. Rice is cooled in the fall and kept below 60°F the rest of the year.

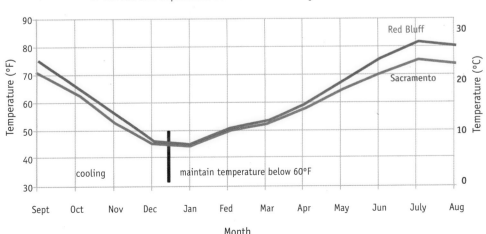

Grain Storage Safety

Grain bins and flat storages are dangerous areas, and people have been killed working in them. Flowing grain can engulf a person standing on it and completely submerse them in only 20 seconds. Wet and moldy grain forms a crust that may remain intact when grain below is removed; someone walking on the crust may break through, become covered in grain, and suffocate. Wet grain can also develop a nearly vertical grain wall during unloading. Further unloading may dislodge the grain wall and engulf people nearby. Moldy grain causes low oxygen and high carbon dioxide concentrations that can suffocate a worker entering a storage area. Moldy grain also produces allergens that can cause severe allergic reactions and permanent lung damage. Dust explosions have destroyed entire grain-handling facilities. A formal safety program will save lives as well as prevent needless damage to equipment and facilities.

California law requires the following for grain handling facilities (see California Code of Regulations, Title 8, Section 5178):

- The employer must issue written authorization for entering the storage or provide a qualified supervisor during the operation.

- The air space above the grain must be tested for low oxygen levels, toxic gases or agents, and dangerous dust levels before anyone enters a grain storage.

- All electrical and mechanical equipment must be disconnected and prevented from operating (locked out, tagged out, etc.) before anyone enters a grain storage.

- Employees must not walk on grain in order to cause it to flow.

- Employees are not allowed near bridged grain.

- When an engulfment hazard exists, a full-body harness with lifeline or a boatswain's chair is required, and a second employee must be present.

- Trained personnel and equipment must be available for rescue operations.

- The facility must be regularly inspected and kept clean. Dust accumulation must not exceed $\frac{1}{8}$ inch in depth.

- Equipment must be designed and maintained to prevent it from becoming a source of ignition.

- For the full text of the California regulations, including requirements for training, record keeping, and dryer and equipment design and operation, see the California Department of Industrial Relations Web site, http://www.dir.ca.gov/Title8/5178.html.

Common sense dictates the following:

- Wear a dust filter or filter respirator when grain is being moved.

- No smoking near storage areas.

- Do not let children near grain storages.

- Lock entrances to storages to keep out unauthorized persons.

Option 1: Manual Control Based on Outside Air Temperature

This method requires information only on local air temperature and can be implemented with relatively little management time. It causes more moisture loss than the other options and may have higher electricity costs (although energy costs are relatively small with any of the options).

Fall and winter

Most rice in California is dried in a column dryer to 16 to 18 percent moisture content, and final drying is accomplished with natural air in bulk storage. The main harvest period is from mid-September through October, when air temperatures are warm and little rain falls, and drying is completed before the end of November. Rice is often fairly cool after reaching the final moisture content through natural-air drying. After drying is finished, aeration fans should be operated to accomplish one complete aeration cycle per month. For example, a fan with a capacity of 1 cubic foot per minute per hundredweight would require about 25 hours of operation per month. Fans do not need to be operated continuously, but the total fan operation time during the previous 30 days should be close to the number of hours needed to produce a complete aeration cycle. Operate the fans when the average air temperature is about equal to the average air temperature for the month (see fig. 8.4). Do not operate fans during a period of high humidity, such as during a rainstorm, or during a period of low humidity, such as during dry north winds.

Operate fans as needed to control any wet spots that may develop or to cool an area that begins to heat up. Under these circumstances the fans should be operated continuously until all signs of moisture or heating are eliminated.

If rice has not been completely dried to a moisture content suitable for milling or sale, usually 14 percent, operate fans whenever ambient conditions are suitable for drying, as described in chapter 6, "Bin and Warehouse Drying."

Spring and summer

Rice should be completely dried before the warm temperatures of spring and summer arrive. Continue to operate fans to provide one aeration cycle per month, but keep rice temperature below 60°F as long as possible. Operate aeration fans when the air temperature is below 60°F. Even during the warmest months, July and August, there are usually periods with low enough air temperature and relative humidity. Under most air conditions rice can be cooled to several degrees below the temperature of the aeration air because of evaporative cooling (moisture loss; see chapter 2). Grains have been successfully stored for a month without aeration, but the risk of hot spots increases as time without aeration increases. In the summer hot spots are located on the top of the storage volume or near south- or southwest-facing exterior walls and are caused by solar radiation and high ambient air temperatures. Ventilate the headspace above the grain to minimize heating during the day.

Option 2: Manual Control Based on Outside Air Temperature and Relative Humidity

This method is based on operating the aeration fans when the air temperature is close to the monthly average and when relative humidity is close to the equilibrium relative humidity corresponding to the desired moisture of the rice. It is more time-consuming to manage than Option 1, but it will minimize moisture loss. During the warmer months of the year, aeration fans may need to operate in the early-morning hours before the workday begins. Option 2 can be implemented using the automated aeration controllers described in Option 3.

After rice has been dried to the desired moisture content in the fall, run aeration fans when the outside air is within a few degrees Fahrenheit of the average air temperature for the month. The average air temperature can be based on the historical data in figure 8.4, or it can be the average temperature for the previous 3 or 4 weeks. Restrict fan operation to periods when the outside relative humidity is in a range from

an equilibrium relative humidity corresponding to the desired rice moisture content to 8 percentage points of relative humidity greater (see table 6. 3, page 76). Fans should be operated long enough to produce one aeration cycle per month. If weather conditions do not allow this much fan operation, the relative humidity range can be expanded by 3 percentage points.

In the late spring, when the average air temperature exceeds 60°F, the threshold for insect activity, aeration fans should be operated only when the outside air temperature is less than this threshold. As the storage season progresses into summer, periods that meet both temperature and relative humidity requirements become less likely. In order to meet the required fan operation time during the month, increase the relative humidity or temperature ranges.

Option 3: Aeration Controllers

Aeration controllers are available that automatically start and stop fans based on desired rice moisture, air temperature, and air relative humidity. They usually have several choices of control strategies and can be user-adapted to meet local needs. Some controllers can be programmed to minimize electricity cost by not operating fans during peak rate periods, and many can interface with temperature probes to activate fans when grain heating is detected. Automated controllers ensure that fans are operated whenever air temperature and humidity conditions are correct for aeration. This is especially valuable when best aeration times are after regular working hours. They tend to keep rice at more consistent moisture and temperature conditions and minimize moisture loss compared with manual control.

Like all automated devices the controllers must be monitored and their sensors calibrated on a regular basis. Automated controllers do not eliminate the need for regular inspection of the stored rice.

Temperature Measurement

Warehouse storages and large bins without stirrers are usually fitted with a network of temperature sensors. These detect a number of problems and simplify aeration management. For example, a localized temperature increase indicates a wet spot from a leak or points to an insect or mold problem. Sensors also allow an operator to determine the level of temperature uniformity in the stored rice and also to determine when aeration is complete. However, rough rice is a good insulator, and the sensors detect temperature rise only within a limited distance. Rice must still be inspected regularly.

Temperature sensors can be built into cables that are suspended from the ceiling or roof. Cables should be spaced about every 20 to 30 feet. Cylindrical bins greater than 40 feet in diameter usually have three cables. The cables and their support must be strong enough to withstand the considerable downward force of rice as it settles or flows past them during unloading. Temperatures are usually read out on a monitor in the storage control room.

Rice Inspection

During storage, inspect rice twice a month when the outside temperature is less than 60°F and weekly when it is warmer than 60°F. Test the discharge air when fans are first turned on and check for off odors that generally indicate rice spoilage. Inspect the surface rice after a major storm to check for leaks and wet rice. If a leak has occurred, take immediate steps to seal the leak. Spread the wet rice and turn on aeration fans or remove the wet rice and dry it. When dry, send sample(s) for rice quality appraisal before mixing it with higher-quality rice.

Elevated grain temperature can indicate fungal growth, but some types of mold can cause grain damage with little noticeable heating. A limited amount of research has indicated that an elevated level of carbon dioxide in a grain mass is a good indicator of excessive microbial activity.

A fast way to check that rice is at a safe moisture content is to measure the relative humidity of the air in the rice. Some electronic humidity probes can be inserted into the rice and will read a constant humidity in about 5 minutes. Permanently mounted probes are being developed. Check humidity in the areas where the rice is likely to have high moisture. Once the relative humidity and air temperature are known, use the equilibrium moisture graph (see fig. 8.1) to estimate the rice moisture content.

Table 8.2. Static pressure that an aeration fan must produce in order to provide an airflow of 0.2 cfm/cwt; the last column indicates the amount of air heating caused by a fan blowing air into a bin

Rice depth	Static pressure	Air temperature rise in positive pressure system
(ft)	(inches w.c.)	(°F)
20	1.1	0.7–1.0
30	1.7	1.1–1.6
40	2.7	1.7–2.5
50	4.0	2.5–3.7
60	5.8	4.0–5.3
70	8.0	5.0–7.4
80	10.5	6.5–9.7

Aeration System Design

Airflow

Aeration requires a minimum airflow of 0.2 cubic feet per minute per hundredweight for bin storage and 0.4 cubic feet per minute per hundredweight for flat storage. These airflow rates cause an aeration front to move through a batch of rice in about 120 hours of fan operation. Most rice storage facilities in California have higher airflow rates because the fan systems are also used for natural-air drying. Table 8.1 shows approximate fan operation times needed for a complete aeration cycle at various airflow rates. A running time meter connected to a fan motor is a convenient way to keep track of fan operation time. The static pressure needed to produce an airflow of 0.2 cubic feet per minute per hundredweight is given in table 8.2. If a fan does not produce a pressure close to that in the table, it is probably not producing adequate airflow and aeration times will be longer than expected. Measuring grain temperature during an aeration cycle and determining when the aeration front progresses completely through the grain mass is the best way to determine the fan operation time needed for a complete aeration cycle.

Heating Caused by Bottom Air Supply

Fans set up to supply air to the bottom of a storage produce a positive air pressure at the bottom of the rice mass. Pressurizing air causes it to heat, and heating air reduces its relative humidity. The heating caused by the fans can overdry the rice, and it also changes the time during the day when the outside air temperature is suitable for aeration. In most on-farm bins and flat storages this heating effect is small (see table 8.2). However, aerating an 80-foot-tall bin can raise air temperature by 10°F and can cause significant overdrying. Commercial storages with bottom-air-supplied fan systems have overdried rice to 10 to 11 percent moisture content instead of the 13 to 14 percent moisture content desired. This heating and overdrying can be prevented in two ways. If the bin roof is strong enough, aeration fans can be installed on the roof, which eliminates the pressure-heating effect. A disadvantage is that roof-mounted fans can cause condensation on the underside of the roof unless outside air is allowed to enter the headspace above the rice during aeration. Another option is to operate fans only when the outside air is very cool and humid, so that the heating effect produces the desired air temperature and humidity conditions for aeration.

Minimizing Uneven Airflow Distribution

Equipment design and management practices can cause uneven distribution of air in stored rice. Rice in areas with little airflow is subject to mold growth and insect activity. Air tends to travel through rice on the shortest path or on the path with least resistance. For example, in flat storage warehouses with tunnel-type air ducts, areas near the floor and between the ducts get the least airflow (fig. 6.9, p. 82). Flat storages are designed with twice the aeration capacity of bins to help compensate for their uneven airflow. The spacing

between ducts should be no greater than the height of the rice (see fig. 6.10, p. 82). Airflow is very uneven in a warehouse when rice is stored at less than eve height. Peaking rice in a bin causes poor airflow through the top rice in the center of the bin (fig. 8.5). If grain must be stored in the peak of a bin, remove the peaked rice as soon as possible. The best airflow uniformity in storage can be gained by storing rice with a level top surface and in a bin with a perforated floor.

Fines in the rice restrict airflow. They tend to collect especially under the fill spout. Spreaders reduce this problem, although they can cause the grain to pack, reducing airflow. In a circular bin, fines tend to collect in the center of the rice. To correct this, unload the center cone of rice after filling the bin, pass it through a scalper and aspirator, and return it to the bin.

Cost of Aeration and Cooling

The direct costs of aeration are the cost of purchasing electricity for fan operation and the cost of moisture lost from the grain. In a typical flat storage with fan capacity equal to 1 cubic foot per minute per hundredweight and a total electricity cost equal to $0.15 per kilowatt-hour, an aeration cycle will cost much less than $0.01 per hundredweight. Aeration can cause a moisture content loss up to 0.3 percent for 10°F of cooling, particularly if aeration is manually controlled. If the rice is already at the desired moisture content before aeration, weight loss caused by aeration will be the greatest cost of aeration.

Refrigerated Storage

Japanese consumers prefer rice that has been stored at 14.5 to 15 percent moisture. This is too high for safe storage at the temperatures encountered in the late spring and summer in California. In Japan, this rice is stored in refrigerated, insulated facilities. The temperature is maintained in the range of 55° to 59°F and relative humidity is controlled at 70 to 75 percent.

References

ASAE (American Society of Agricultural Engineers). 1995. Moisture relationships of plant-based agricultural products. Standard D245.5. St. Joseph, MI: ASAE.

Multon, J. L., ed. 1988. Preservation and storage of grains, seeds, and their products. New York: Lavoisier.

Navarro, S., and R. Noyes, eds. 2001. The mechanics and physics of modern grain aeration management. Boca Raton: CRC.

Reed, C. R. 2006. Managing stored grain to preserve quality and value. St. Paul American Association of Cereal Chemists International.

Figure 8.5. Effect on airflow distribution of not leveling the grain.

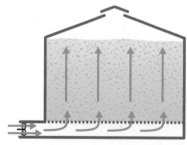

Leveled rice produces uniform airflow in a bin with a perforated floor

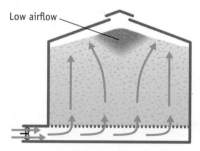

Low airflow

Peaked rice causes a region with low airflow

CHAPTER 9

Michael Poe

Controlling Spoilage: Insects and Decay in Stored Rice

Prevent Spoilage

- ⊚ Clean storage area and bins before harvest.

- ⊚ Apply insecticides to storage areas before harvest.

- ⊚ Clean rice before storage.

- ⊚ Do not mix old crop and new crop rice.

- ⊚ Measure the rice moisture content as it is placed in storage.

- ⊚ Do not allow fines to collect in a small area in the storage.

- ⊚ Aerate rice to keep it at a uniform moisture content below 60°F as long as possible.

- ⊚ Regularly measure rice temperature and moisture content.

- ⊚ Regularly inspect for live insects, sprouting rice, wet spots, and off odors.

- ⊚ Control insect outbreaks using registered insecticides or alternative methods.

- ⊚ Cool hot spots with aeration or stirrers with fans.

Environmental Factors Affecting Insects

Among the multitude of known insect species, only a tiny minority are stored grain dwellers. For example, of the thousands of identified members of the weevil family *(Curculionidae)*, just three species of the genus *Sitophilus* are major stored grain pests, and only one species, the rice weevil, is typically found in stored rice in California. Stored grain insects are tolerant of extreme drought and are uniquely capable of exploiting the artificial environment of a grain elevator, with its concentration of an abundant food supply in a compact, favorable habitat. Without imposed constraints, stored grain is an ideal environment for an explosion in insect populations.

Temperature

Temperature controls the rate at which insects multiply and infest rice. Most stored rice insects originate from subtropical areas; they do not feed or reproduce at temperatures below 60°F, and their growth is severely limited by temperatures above 100°F. Outside of the optimal range, their reproductive processes are disrupted. Both the rice weevil and the Angoumois moth exhibit a similar lower and upper temperature limit for successful reproduction (table 9.1). The rice weevil, however, is the least tolerant of low relative humidity (requiring humidity greater than 60 percent) compared with the other major stored rice pests, which can survive humidity as low as 30 percent. The lesser grain borer and confused flour beetle require grain temperature about 70°F to increase populations. As a general rule, rice at temperatures between 70° and 90°F is at risk of severe insect damage.

Within the range of temperatures required for successful reproduction there is a dramatic difference in the potential for population growth.

At a grain moisture content of 14 percent, an initial population of 50 mated pairs of rice weevils produced 14,000 insects within 5 months at 80°F (fig. 9.1). At the minimum temperature (60°F), the insect population increased to only 1,000 in the same period. Rice weevils require less time to complete their life cycle than do lesser grain borers across a range of temperatures (fig. 9.2). At around 80°F the time required to complete the life cycle is about 1 month for the rice weevil and 2 months for the lesser grain borer. In contrast, at 60°F, it takes 125 days for the rice weevil and 175 days for the lesser grain borer to grow from an egg to an adult. Keeping the rice cool throughout the grain mass is fundamental to insect population control.

Moisture

Insects living in stored rice depend on grain moisture to sustain life. The optimal rice moisture

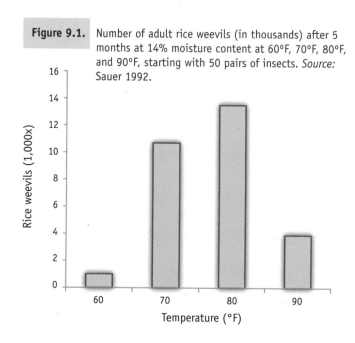

Figure 9.1. Number of adult rice weevils (in thousands) after 5 months at 14% moisture content at 60°F, 70°F, 80°F, and 90°F, starting with 50 pairs of insects. *Source:* Sauer 1992.

Table 9.1. Approximate minimum and optimum temperature and relative humidity (RH) for storage insect population increase

Insect	Temperature, F°		Minimum RH (%)
	Minimum	Optimum	
rice weevil	63	81–88	60
Angoumois moth	61	80–86	30
lesser grain borer	73	90–95	30
confused flour beetle	70	86–91	1

Source: Sauer 1992.

range for insect growth is narrow. The rice weevil, for example, is unable to reproduce below a grain moisture content of 10 percent, and its reproduction rates are maximized at 14 percent (fig. 9.3). The long-term survival of the rice weevil diminishes with decreasing grain moisture content (fig. 9.4). Storing rice at a low moisture content to control insects is an option only for seed rice; such low moisture would put milled rice at risk of fissuring, poor milling yields, and lost weight at the time of sale. Although some insects can successfully develop in grain moisture contents above 18 percent, fungal growth becomes a more critical concern at this point. In commercial practice, rice moisture content is not used to control insects, leaving temperature management and sanitation as the primary nonchemical tools for preventing infestation.

Mobility

Some stored rice insects are strong fliers, including the lesser grain borer and the Angoumois moth. These insects can travel several miles in search of food sources. Nonflying storage insects are capable of moving between bins in a short period of time. Stored grain insects have a tendency to move frequently, sometimes in mass. Conditions that promote insect movement include large populations in a small amount of grain, grain disturbance, or changes in grain temperature.

Stored grain insects can be very sensitive to small changes in grain temperature. Research has shown that insects will quickly migrate to warm spots in the grain, such as areas created by microbial activity,

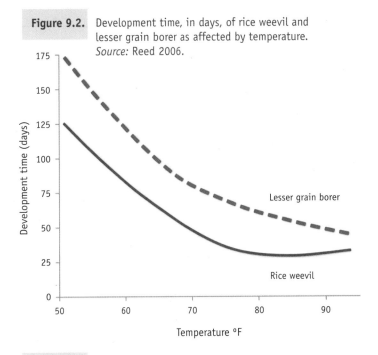

Figure 9.2. Development time, in days, of rice weevil and lesser grain borer as affected by temperature. *Source:* Reed 2006.

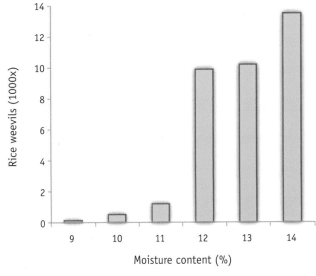

Figure 9.3. Number of adult rice weevils (in thousands) after 5 months at 80°F at selected grain moisture contents, starting with 50 pairs of insects. *Source:* Sauer 1992.

Figure 9.4. Survival of rice weevil adults at 85°F in wheat at selected moisture contents. *Source:* Cotton et al. 1960.

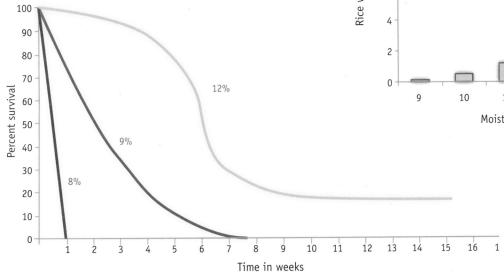

the addition of high-moisture-content rice into a bin of dry rice, or seasonal temperatures. Insects flee areas of high temperature. Heating bins prior to filling is an effective nonchemical method for reducing insect levels.

Growth and Development of Rice Storage Insects

The sudden appearance of large numbers of insects is caused by the female insect's ability to lay a large number of eggs. Depending on the species and environmental conditions, one female may produce a dozen to over 1,000 eggs. A few days after the eggs are laid, they hatch into larvae. The larvae develop through discrete stages called larval instars. Larvae actively feed at all instar stages. At the end of each instar, the larva develops a new exoskeleton and sheds the old one, allowing it to increase in size. Depending on the species, larvae go through three to six instars. After the final larval instar, insects enter the pupal stage, where metamorphosis is completed. The pupa are no longer active and do not feed as the insect transforms into an adult. The pupating period varies between species but is generally complete within a few weeks to several months. Once the adult insect emerges, it is capable of reproducing within a few days and the cycle repeats. The complete life cycle lasts from a few weeks to many months depending on the species and environmental conditions.

Major Insects in Rice

Stored grain insects are classified into four categories based on their potential to cause damage. Major pests are those responsible for most of the damage and are particularly well adapted for survival in stored grains. The major pests are divided into two subgroups: internal infesters and external infesters. Minor pests are a larger group of insects capable of significant grain damage if the appropriate environmental conditions are met, such as high grain moisture. Incidental insects such as grasshoppers and flies can find their way into the bins, but they are generally of little concern.

FGIS Standards: Insect-Infested Paddy Rice

Rice is considered to be infested and supplemental grades are applicable when threshold levels of insects injurious to stored rice are present.

Representative sample from an incoming load:

- 2 or more live weevils, or
- 1 live weevil and 1 or more other live insects, or
- 5 or more other live insects

A representative sample should reflect the condition of the entire load. A representative sample may be a composite of random samples if the grain is well mixed. If the grain is stratified (e.g., in different temperatures), each area should be sampled and analyzed separately. More information can be found at the USDA-GIPSA Web site, http://www.gipsa.usda.gov.

Lot sample from stationary storage:

- 2 or more live weevils, or
- 1 live weevil and 1 or more other live insects, or
- 5 or more other live insects, or
- 15 or more live Angoumois moths or other live moths found in, on, or about the lot.

Areas within the grain mass that are 10°F or warmer than the average bin-wide temperature should be intensively sampled. The grain adjacent to the south-facing walls of the bin or silo will warm the quickest in the spring.

Loading/unloading continuous sample:

The minimum sample size is 1.1 pound (500 g) per each 100,000 pounds of rice.

- 2 or more live weevils, or
- 1 live weevil and 1 or more other live insects, or
- 5 or more other live insects.

Grain Bin Sampling Report

Inspector's name_____ Date_____

Bin number _____ Bin diameter _____ Bin wall height_____

Sample Information

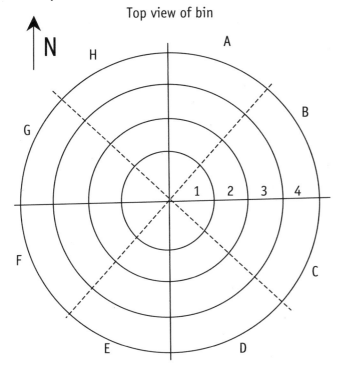

Top view of bin

N

H   A

B

G

F

1 2 3 4

C

E   D

(H) Hatch location (draw)

X   Sample locations (draw)

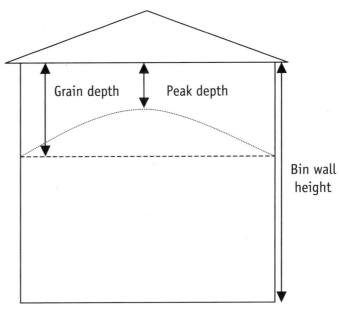

Sample/grain depth

Grain depth   Peak depth

Bin wall height

Measure from the top of the bin

Grain depth _____

Peak depth _____

Sample location e.g., A-1	Depth of probe	Can no.	% MC	Temp	Insect count

Internal Infesters

Three insect species lay eggs or develop within the kernels of stored rice. These insects damage stored rice not only by feeding, but also by the heat, moisture, and waste products produced by their activity.

Rice weevil

The adult rice weevil *(Sitophilus oryzae)* has a long snout, which is tipped with chewing mouthparts, and elbowed antennae (fig. 9.5A). The rice weevil is sometimes confused with the granary weevil *(Sitophilus granus)*. However, the granary beetle is a major pest of stored grain in upper Midwest and cold climates of the United States and is typically not found in rice stored in California. The rice weevil has two light-colored areas on the wing covers and round, closely spaced pits on the thorax. The granary weevil is uniformly colored and the pits on the thorax are elliptical and widely spaced. Both are dark brown and about ⅛ inch in length (fig. 9.5B). The rice weevil is thought to be capable of flight, while the granary weevil is not. It can be transported in infested lots of rice grain.

The egg, larval, and pupae stages of the rice weevil are spent inside the rice kernel. Generally there is only one larva per seed. Once mated, the female chews a hole in the kernel with her long snout and excavates a small cavity into which she places an egg. She then seals the hole with a gelatinous plug. To the naked eye there is little evidence that the kernel has been damaged. An adult female can lay about 400 eggs in her lifetime. Eggs hatch in few days under favorable temperature and moisture conditions. The time from egg to adult is only 35 to 40 days; the larva and pupa remain inside the kernel. The adult chews out of the kernel, leaving behind the characteristic weevil damage (fig. 9.5C). Adult weevils live for 5 to 8 months.

Rice weevils prefer cooler temperatures and higher grain moisture than do the lesser grain borers (below). If large rice weevil infestations are allowed to develop, they produce large amounts of heat and moisture, which may initiate grain heating that encourages mold growth.

Lesser grain borer

The lesser grain borer *(Rhyzopertha dominica)* can be one of the most damaging insects in stored grains (fig. 9.6A) and prefers a hot, dry environment. They are slow-moving insects that remain localized. Lesser grain borers consume large amounts of grain, and infestations produce grain dust and a sweet, musty odor. The dust may result in compacted areas within the grain that lead to zones of poor aeration or poor fumigant penetration. Infestations increase the potential for dry grain heating. Mold growth is stimulated by the heat and moisture that the insects produce.

The body is cylindrical and about ⅛ inch long. The thorax forms a distinctive helmetlike hood over the head. The wing covers have rows of indentations; the adult is a strong flyer. Most larvae develop inside the rice kernel and are

Figure 9.5. Adult rice weevil (A), adult on long grain rice (B), and exit holes on rice kernels (C). *Sources:* Degesch America (A); Joseph Burger, Bugwood. org (B); Clemson University/ USDA Cooperative Extension Slide Series, Bugwood. org (C).

A   B   C

Figure 9.6. Adult lesser grain borer (A), larva inside a grain kernel (B), and exit holes in infested grain (C). *Sources:* Degesch America (A); USDA/ARS.GMPRC (B); Clemson University/USDA Cooperative Extension Slide Series, Bugwood.org (C).

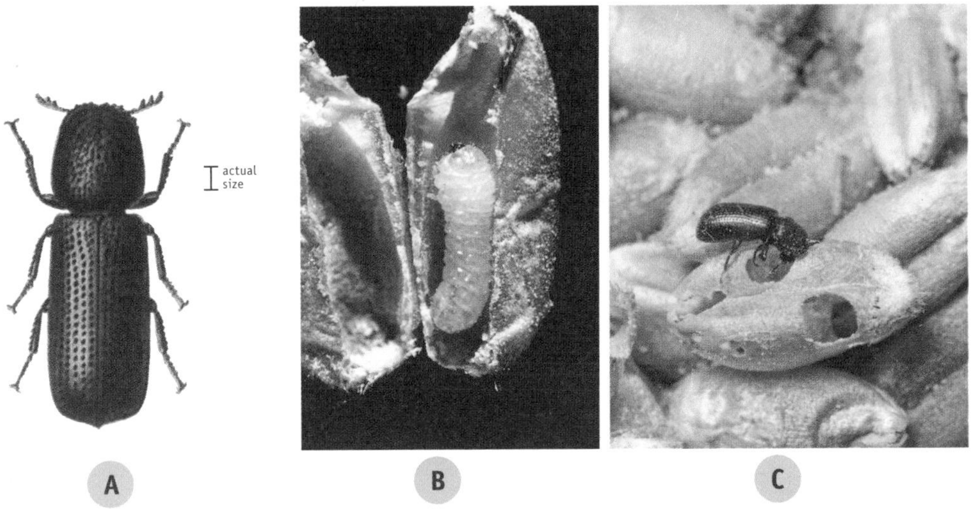

A   B   C

Figure 9.7. Adult Angoumois grain moth (A), and larvae-infested grain (B). *Sources:* Degesch America (A); Clemson University/USDA Cooperative Extension Slide Series, Bugwood.org (B).

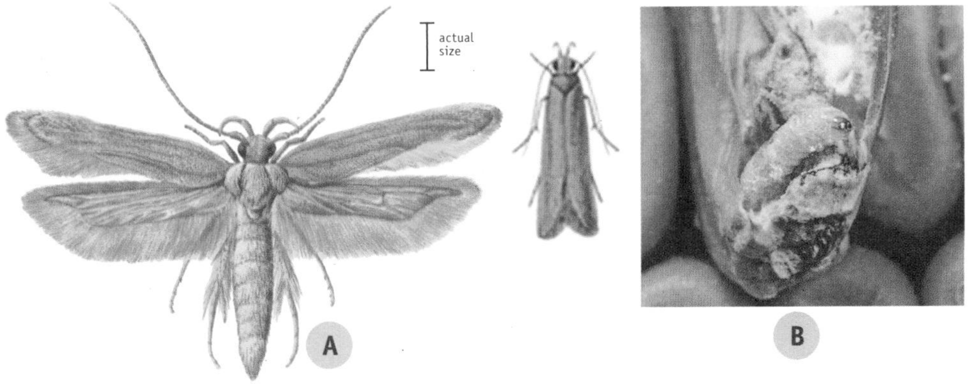

A   B

invisible during most of their life cycle. The body of the larva is grublike (fig. 9.6B). Females lay 300 to 400 eggs, placing them among the grains rather than inside the kernels. The pear-shaped eggs hatch into larvae in 6 to 10 days. The larvae immediately burrow into the kernels or penetrate through cracks in the seed. Adults emerge after a few days (fig. 9.6C) and live an average of 7 to 8 months.

Angoumois grain moth

The Angoumois grain moth *(Sitotroga carealella)* (fig. 9.7A) does not cause as much damage as the rice weevil or lesser grain borer. Nevertheless, at warm temperatures it can cause major grain deterioration. It requires grain masses that are easily penetrated, because the adult has a delicate body; consequently, it is not found deep in the grain mass. Adult moths tend to flutter over the surface of the grain, in contrast to other moths that tend to fly to vertical surfaces looking for

mates. The adult moth is ¼ inch long. Its buff-colored wings with two dark spots are folded back over the body, forming a distinct arrowhead-shaped pattern. The wing tips are fringed with long hairs.

Females lay about 80 eggs at a time among the kernels. Newly hatched larvae have three tiny pairs of legs that become less apparent in later stages. They enter the kernel by chewing or entering through cracks (fig. 9.7B). Larvae maintain a C shape while inside the kernel. Only the larval stages feed on the grains. Before pupating, the last instar larva chews an "escape window" for the adult. The nonfeeding adult easily emerges from the damaged kernel.

External Infesters

Two main pests feed on the outside of rice kernels, preferring the germ (embryo), strongly compromising rice germination. External infesters spend their entire life cycle outside the kernel and are not a major concern in stored rice.

Confused flour beetle

An occasional problem in processed and packaged forms of rice is the confused flour beetle (*Tribolium confusum*, fig. 9.8A). Members of this family of insects are considered foragers, feeding on decaying grain dust, decaying plant material, and molds. Confused flour beetles do not damage dry grain, but they can become numerous in the grain mass and create warm, moist conditions conducive to mold growth. The adult confused flour beetle is ⅛ inch long, shiny, and reddish brown (fig. 9.8B). Larvae are light tan with a darkened head and a darkened forked process at the tip of the abdomen. Larvae have three pairs of legs on the body segments immediately behind the head.

Adult females lay 400 to 500 eggs over a period of several months. Eggs hatch in 3 to 5 days, and larvae go through a series of instars while feeding on grains. Fines such as broken kernels and grain dust provide a good environment for larval development, allowing the confused flour beetle to survive and reproduce in rice with a moisture content as low as 8 percent. The development from egg to adult takes about 26 days, and adults live for about 1 year. When confused flour beetles are overcrowded or agitated, they secrete chemicals called quinines that have a pungent, penetrating odor and cause heavily infested grain to turn pink.

Grain mite

The grain mite (*Acarus siro*, fig. 9.9A), also known as the flour mite, is in the same class as spiders, rather than insects. It often feeds on fungi associated with stored grain, such as *Aspergillus flavus*. Thus, grain mites can be particularly problematic when moist grain (greater than 14 percent moisture content) is held for long periods at cool temperatures (50°F). Grain mites grow through a greater number of development stages than do insects, and in general they grow more rapidly. They are wingless and microscopic in size (fig. 9.9B). Once hatched, the mite larvae quickly molt into nymphs. After the first nymphal instar, the mite can develop into an adult in less than 2 weeks; however, the nymph may assume a dormant form (hypopus) if environmental conditions are not favorable for it to develop into an adult. Favorable moisture (75 to 85% relative humidity) is more critical than temperature for it to complete its life cycle. Mites can complete their development at temperatures as low as 45°F, with an optimal temperature around 75°F. As a result, mites are often found in damp grain at cool temperatures. Mites feed on embryos of the seed or on molds infesting the grain; their presence indicates high-moisture conditions (fig. 9.9C). A mite infestation kills seeds, reduces germination, and reduces the milling quality of grains. Mite-infested rice can develop a mintlike odor. Fumigation may be ineffective because of the unusual ability some mites have to survive under adverse conditions by entering a resting stage.

Figure 9.8. Adult confused flour beetle (A) and adults on kernel of wheat (B). *Sources:* Degesch America (A); Clemson University/USDA Cooperative Extension Slide Series, Bugwood.org (B).

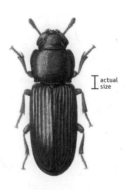

A

B

Figure 9.9. Adult grain mite (A), mite-infested wheat germ (B), and mite-damaged grain (C). *Sources:* Degesch America (A); David Crossley, Central Science Laboratory, UK (B, C).

Figure 9.10. Adult sawtoothed grain beetle (A), red flour beetle (B), and Indian meal moth (C). Relative size is not to scale. *Source:* Degesch America.

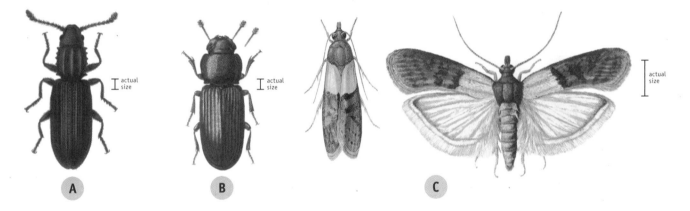

Minor Insect Pests of Stored Rice

Other insects that can occasionally be found in stored rice and rice products in California are the sawtoothed grain beetle, red flour beetle, and Indian meal moth (fig. 9.10).

The adult sawtoothed grain beetle (*Oryzaephilus surinamensis,* Latin for "rice lover") is about 1/8 inch long and dark brown. Diagnostic features include a flat body and six prominent teeth along each side, with three longitudinal ridges on the top of the thorax. The sawtoothed grain beetle rarely flies and usually lives for 6 to 10 months. Female sawtoothed grain beetles usually emerge in March or April and lay an average of 300 eggs among loose grains. Egg laying begins about 5 days after emergence and

continues for 3 to 4 weeks. The life cycle can be completed in 51 days or less depending on the temperature. There may be as many as 6 to 7 generations under warm conditions of 85° to 95°F and a 70 percent relative humidity, with fewer generations throughout the winter months. The larvae are active and feed on damaged grains. Undamaged dry rice kernels are not susceptible to predation; high-moisture rice can be at risk.

The red flour beetle (*Tribolium castaneum*) is similar to the sawtoothed grain beetle in behavior and the range of food products infested. Infested grain emits a foul odor and has a moldy flavor when cooked. The adults feed on grain dust and broken kernels but not on undamaged whole grain kernels. The adults are about 1/8

inch long and are flat, shiny, reddish-brown, and elongated. The adult is a strong flier. It is similar in appearance to the confused flour beetle (see fig. 9.8). The last segments at the tip of the antennae of the red flour beetle are abruptly larger than the preceding ones, forming a three-segmented club. In contrast, the antennae segments of the confused flour beetle increase in size gradually from the base to the tip. Female beetles lay 300 to 400 eggs in flour or broken grain dust over a period of 5 to 8 months (two to three eggs per day) when the temperature is over 68°F. Eggs hatch in 5 to 12 days, and the larvae feed both outside and inside the grain. The life cycle from egg to adult takes 15 to 20 days under optimal conditions, which are a temperature of 95°F and relative humidity of 10 percent or less. The red flour beetle cannot complete development under 68°F. It will fly when the temperature is 77°F or higher, so infestations can spread quickly. The red flour beetle is more common in warmer parts of the southern United States.

The Indian meal moth *(Plodia interpunctellar)* can infest both stored grain and cereal products. It generally stays within 18 inches of the surface of the grain. Consequently, Indian meal moths become active in the spring before beetles and weevils that may have penetrated deeper into the grain mass. Adult moths are about $\frac{3}{8}$ inch long with a wingspread of about $\frac{5}{8}$ inch. The wing tips appear reddish-brown while the bottom two-thirds of the wing is light gray. The adult lives for about 14 days. Temperature, relative humidity, and food moisture content has a significant effect on insect development times. Adult moths usually emerge, mate, and lay eggs at night. Indian meal moth breeds and develops within a temperature range of 59° to 95°F and from 25 to 95 percent relative humidity. Females lay from 40 to 400 eggs in 18 days on or adjacent to food material. Eggs start hatching within 4 to 8 days. The optimal development time of 30 days is found at 86°F and 75 percent relative humidity. Larvae feed on both external and internal parts of the rice, especially the embryo. Larvae spin silk to bind grains together and then spin a web over the grains. Undisturbed surface grain with heavy infestations may become completely webbed over, underneath which the larvae are partially protected from changes in their environment. This webbing prevents uniform grain aeration and impedes effective fumigation. An exposure of 2 to 3 days to temperatures of 5°F or lower kills the more susceptible stages (larvae and adults), but eggs require longer to kill (3 weeks).

Environmental Factors Affecting Microorganism Growth

The growth of microbes (mainly molds and bacteria) is influenced by environmental conditions, the most important of which are temperature and humidity. The concept of water activity (see chapter 10) is related to the amount of water available for microbial growth. Small changes in water activity and temperature can have desirable or undesirable consequences for grain integrity. Under the right conditions, harmful microbial growth can occur in just 7 days, and this growth is influenced dramatically by water activity (table 9.2). The visual presence of a fungal mass is the result of prolonged growth. At water activity below about 0.65, fungal growth will not occur regardless of the temperature. Above 0.65, the optimal water

Rice damaged by storage mold. *Photo:* Michael Poe.

Figure 9.11. Relationship of grain temperature, water activity, and mite, mold, and insect development.

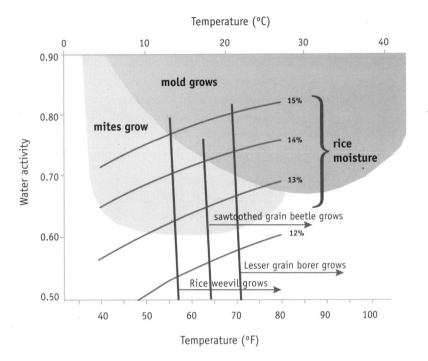

nutrients needed for growth. Fungi reproduce primarily by means of small, light, airborne spores that are easily distributed by the wind. Spores germinate and grow whenever moisture and temperature conditions are favorable.

Field fungi

Rice growing in the field can be contaminated by numerous fungi. Under high moisture and high temperature conditions, some fungi, such as common smut, may develop over wide areas before harvest. The main field fungi (sometimes referred to as field molds) require an equilibrium relative humidity of 95 to 100 percent, typically associated with rice with a moisture content of 25 percent or greater. Field fungi can be extremely detrimental to rice quality if the freshly harvested rice is not adequately aerated. After the rice is harvested and the rice is dried, the field fungi die and a different group of fungi, the storage fungi, develop.

The production of carbon dioxide and ethanol in a grain mass is indicative of microbial activity (fig. 9.12). Ethanol and carbon dioxide are

activity for growth depends on temperature. Above or below the 85°F temperature for optimal mold growth, the minimum water activity required for growth increases (fig. 9.11).

Bacteria

The active growth of bacteria found on seeds generally requires a rice moisture content in equilibrium with 0.92 to 0.95 water activity. In rough rice this is a moisture level above 26 percent, a level that is not usually found in storage bins. As a result, bacteria are not involved in storage losses other than in very wet grains or in the final stages of microbiological heating. However, bacterial decay of undried rice before it is ready for long-term storage can lead to rapid development of off odors and off flavors.

Fungi

Fungi (molds) are multicellular organisms without roots, leaves, or chlorophyll. Therefore, they must live off other materials, including grains. The vegetative parts of fungi produce enzymes that interact with grain tissue to extract the

Table 9.2. Days before visible appearance of fungi on rice at 77°F

Water activity	Number of days	
	Rice	Glutinous rice
0.98	7	7
0.95	9	9
0.90	10	10
0.85	10	10
0.80	13	17
0.75	20	19
0.65	57	73

Source: Abdullah and Othman 2000.

Figure 9.12. The evolution of ethanol (A) and carbon dioxide (B) over time in Akitakomachi rice at selected moisture contents. Rice was harvested at a range of moisture contents and immediately placed in sealed containers under laboratory conditions. *Source:* Champagne et al. 2004.

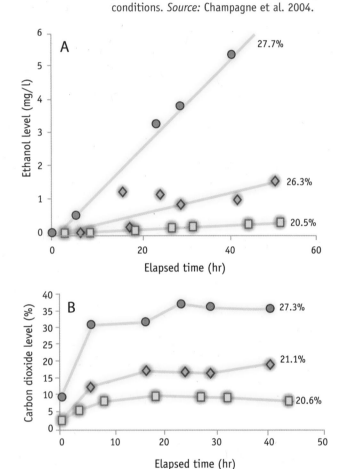

(fig. 9.13). Such deterioration in quality can occur when rice above 26 percent moisture content is not aerated within a few hours of harvest. Timely transport of the freshly harvested rice to the dryer minimizes the development of off odors. Generally, rice should not be harvested above a moisture content of about 24 percent because typical commercial handling time allows off odors to develop between harvest and the start of column drying.

Storage fungi

Rice grains are seldom infected with storage fungi in the field. Instead, the infection takes place in transport equipment or at the elevator where the grain is dried and stored. Unlike field fungi, storage fungi grow at an equilibrium relative humidity less than 95 percent. In fact, these fungi can grow and reproduce at a relative humidity as low as 70 percent where there is no free water, and at temperatures down to 23°F. Storage fungi occupy a distinctly different ecological niche than do the field fungi. Therefore, with few exceptions, fungi introduced during postharvest handling cause spoilage of stored rice.

Characteristics of major storage fungi

The common storage fungi are several species of *Aspergillus* and *Penicillium* (table 9.3). Each develops at a different relative humidity. Although expensive for growers to use, commercial laboratories can identify fungi; growers can use this information to estimate the range of rice moisture and temperature conditions in spoiled areas.

associated with rice having off odors and poor taste. If the wet grain is left unaerated, an off odor will develop, which results in a detectable loss of taste quality even in the milled white rice

Table 9.3. Equilibrium moisture content of rice grains at relative humidity of 65 to 90% and fungi likely to be present

Relative humidity (%)	Grain moisture (%)	Fungi
70–75	14.0–15.0	*Aspergillus restrictus, A. glaucus*
75–80	14.5–15.0	*Aspergillus restrictus, A. glaucus, A. candidus*
80–85	16.0–18.0	*Aspergillus restrictus, A. glaucus, A. candidus, A. flavus*
85–90	18.0–20.0	*Penicillium* spp.

Source: Sauer 1992.

Figure 9.13. Sour or silage flavor in white rice obtained from paddy rice at moisture contents of 17–27% and immediately dried or held for 48 hours. Data for M-202 (17, 24%) and Akitakomachi (18, 21, 25, 27%). *Source:* Champagne et al. 2004.

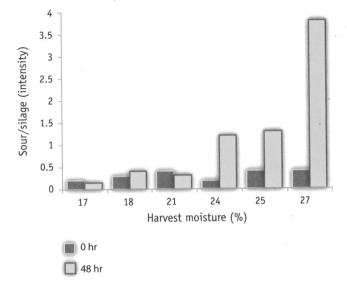

☐ 0 hr
☐ 48 hr

Figure 9.14. Effects of various species of *Aspergillus* on the germination of seeds stored at 86°F and 85% relative humidity. *Source:* Sauer 1992.

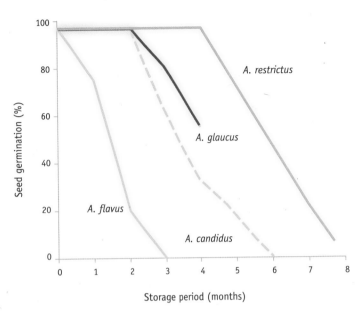

Aspergillus restrictus is often the first fungus to appear, at grain moisture contents of 14 to 14.5 percent. The fungus kills and discolors the germ, giving it a purplish-black color. It grows too slowly to contribute to any detectable rise in temperature. It is responsible for a musty odor in grain. It is closely associated with granary and rice weevils. A high incidence of weevils may indicate that the grain is heavily contaminated with *A. restrictus*.

Aspergillus glaucus, the most common storage fungus, actively grows at moisture contents of 14.5 to 15 percent. It kills and discolors the germ, causing a musty odor. Eventually, the colonization of the rice results in caking that is apparent to the naked eye. If growing vigorously, *A. glaucus* can increase temperature up to 95°F. Most cases of fungus-caused spoilage, especially in the early stages of storage, involve *A. glaucus*. Early stages of *A. glaucus* infestation cannot be detected without a microscope. Its dominance and almost universal presence in stored grain suggest that most mold problems start at grain moisture levels only a little too high for safe storage, rather than at excessive moisture contents.

Aspergillus candidus occurs at moisture contents of 15 to 15.5 percent. *A. candidus*, a major storage fungus, causes discoloration of the entire kernel while turning the germ black. The presence of this fungus is an indication that serious grain deterioration is underway. Its growth results in heating of the total grain mass up to 130°F, at which point the grain is a total loss. A pronounced rise in grain temperature between sampling dates may indicate heightened *A. candidus* activity and is cause for alarm.

Aspergillus flavus develops at moisture contents between 16 and 18 percent. It kills the germ, discolors the seed, and causes rapid heating. If *A. flavus* is detected, major spoilage has already occurred. Under certain conditions it can produce aflatoxin, a very potent poison. If *A. flavus* is discovered, immediate drying and cooling of the grain is essential.

Penicillium spp. are fungi that develop in rice stored at a relatively high moisture content and low temperatures. *Penicillium* can develop at temperatures as low as 23°F. They cause kernel discoloration, caking, and a musty odor. They can also produce mycotoxins.

Effects of storage fungi on seeds

Germination. Seed viability can be reduced before the fungus is visible even with a microscope. Weakening or killing of the seed embryo happens before kernel discoloration. The seed will not germinate if discoloration is easily detected. The speed at which seed viability is lost depends on the storage fungi present (fig. 9.14). *Aspergillus flavus* may kill an entire seed lot within 3 months. In contrast, seed lots infected with *A. restrictus* may not lose all germinability even after 8 months.

Grain heating. Moist grain was once thought to heat because of its own respiration, but fungal activity has been found to be the source of the heat. Heated grain does not become moldy, but moldy grain becomes heated. Heating in stored rice begins with high-moisture grain from incomplete drying after harvest, moisture gain through bin leaks, insect activity, or moisture transfer resulting from temperature gradients in the grain mass. As the moisture increases due to fungal growth, the procession of different species as described above raises the grain temperature up to 130°F and may maintain that temperature for several weeks. The blackened stored grain is a total loss at this point. The process may then subside or continue to the next step, in which thermophilic bacteria and fungi begin to grow. They can raise grain temperature to as high as 167°F, the maximum attainable by microbiological heating. Above this temperature chemical processes can raise the temperature to the point of spontaneous combustion.

Musty Odors, Caking, and Decay

These characteristics, which are detectable by the nose or unaided eye, indicate advanced stages of spoilage. Substantial fungal growth can occur before it becomes readily apparent. Fungal growth can generally be seen on the grains before musty odors can be detected. Caking results from webbing of fine threadlike mycelium between and within the kernels. The caking may sometimes be only a few inches thick, consisting of rotten kernels and fungal mycelium, while the bulk of the grain underneath remains sound. Regardless of the depth, caking represents the final stages of decay.

Managing Stored Rice to Prevent Insects and Fungi

Prevent outbreaks of storage insects by minimizing their numbers before the harvest season begins. Most storage insects and mites are not brought in with the newly harvested crop; they live in or near the storage in pockets of grain or trash. Thorough cleaning reduces initial insect levels by removing the remnants of last year's crop. Sweep or vacuum under bin floors, around the base of the bins, and any surfaces where dust, trash, or rice accumulates. Clean out conveyors and scalping equipment. Insects can also live in grain residue left over in combine and transport equipment. Remove any outside piles of grain, hulls, or trash. Control weeds around the storage area, since they provide excellent habitat for insects and rodents.

Spray the inside and outside surfaces of bins and warehouses with an approved residual insecticide 4 to 6 weeks before filling to control invading insects. Treat all cracks, crevices, areas around doorways, and other places that provide hiding places for insects. Raising the temperature in a bin prior to filling drives insects out.

Store only clean rice. Broken kernels and other fines are attractive to some insects and are more susceptible to mold than are whole kernels. Weed seeds, brokens, and other fines also restrict airflow, fostering high moisture and temperature conditions that favor insect and mold growth. The combine should be adjusted to eliminate the maximum amount of chaff and weed debris. Maintain and adjust the combine, as well as auger conveyors, to minimize broken kernels. Check the adjustments by inspecting combined rice, making sure that there are few or no kernels without hulls. Clean the rice prior to storing to reduce the volume of fines. The material separated from the grain by screening or aspiration is a good environment for insect growth and should regularly be removed from the storage area.

Small particles accumulate beneath the grain spout as the grain fills the bin. This produces a cone of relatively solid grain mass starting at the floor and extending to the top of the grain. This area has low airflow and presents favorable conditions for insect and mold growth. Avoid concentrating fines under the spoutline by using a grain spreader that distributes fines uniformly throughout the grain mass. A disadvantage of spreaders is their tendency to pack the rice and reduce airflow. Grain bins can be filled without a spreader and then the center grain can be emptied and recleaned. Level the rice in storage to maintain uniform airflow through the grain mass. Air follows the path of least resistance, so it will flow through the shallower portion of the grain. Overfilling bins makes leveling difficult and leads to high-moisture rice at the peaks.

Set a limit on the maximum moisture content of the rice entering the storage and monitor the rice to ensure that it does not exceed that moisture content. The maximum moisture content should be low enough so the rice can be dried to a safe level before the outside temperature drops in November (see chapter 8, "Storage and Aeration"). Never mix new-crop rice with last year's crop. Rice stored over the summer is prone to insect damage and may serve as a source of insects for the new crop.

Regularly aerate rice to keep it uniformly cool (see chapter 8). Storing rice at temperatures below 60°F prevents insects from feeding and reproducing (see fig. 9.11). Mold does not grow below a 65 percent equilibrium relative humidity (0.65 water activity) at warm temperatures. At 50°F it does not grow below 75 percent equilibrium relative humidity. Cooling the rice with aeration does not completely eliminate the possibility of insect infestation, but it does provide a high degree of protection, especially when used in conjunction with other pest control strategies.

Monitoring

Regularly measure and record the grain temperature throughout the duration of storage because an insect infestation or mold development usually causes rice to heat. Install

Figure 9.15. Guidelines and sampling areas for insect and mold monitoring.

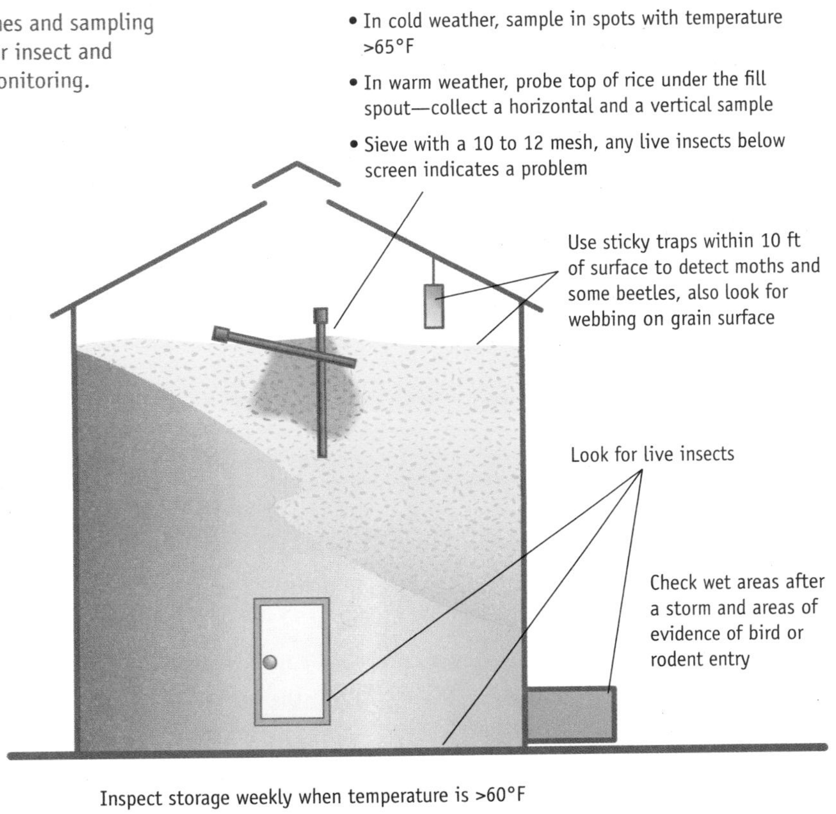

- In cold weather, sample in spots with temperature >65°F
- In warm weather, probe top of rice under the fill spout—collect a horizontal and a vertical sample
- Sieve with a 10 to 12 mesh, any live insects below screen indicates a problem

Use sticky traps within 10 ft of surface to detect moths and some beetles, also look for webbing on grain surface

Look for live insects

Check wet areas after a storm and areas of evidence of bird or rodent entry

Inspect storage weekly when temperature is >60°F

temperature sensors attached to automatic recording equipment prior to filling the bin. Space cables 20 to 30 feet apart, with sensors every 6 feet along the cables. A few degrees of temperature rise in even one sensor signals the need for immediate corrective action.

Grain damage can be prevented by early detection of insect infestation. Inspect the storage twice a month during the winter and when a temperature rise signals an insect or mold problem. Roof leaks cause discolored grain and may cause grain to sprout. Mold causes off odors that can be detected in exhaust air and will be noticeable in a vacuum-probed sample. Mold also darkens the grain and causes it to clump together. Visible signs of insect activity are apparent in kernels as emergence holes or missing embryos, frass, cast skins, webbing, or mold on the rice. A musty or moldy smell can indicate the presence of lesser grain borers. During warm weather, borer infestations generally occur near the surface of the grain, especially in the area below the spout line (fig. 9.15). Several locations in the upper portion of the rice should be sampled with a probe. During cold weather, grain temperature is a good indicator of the areas to sample. Generally, areas with temperature greater than 65°F are likely spots for insect development. In the spring, moths become active when the temperature in the bin approaches 80°F. Dichlorvos strips or pheromone traps are effective monitoring tools. Traps should be suspended within 10 feet of the grain surface. During the summer, check the grain weekly and continue to aerate rice. Insect infestations are usually near the surface of the grain in the summer and fall. In contrast, during cold weather, stored grain insects congregate at the center and lower portions of the grain mass and may escape detection until extensive heating has developed.

Stored rice can be inspected either by observing the grain surface or by collecting samples with a grain probe (e.g., a deep bin cup probe) in the bin center, periphery, and surface locations. Each sample should be carefully examined for signs of insect activity, such as exit holes, removal of the germ, and webbing. Grain screens are also useful. A No. 10 sieve with a 2-millimeter aperture is best for rice. Examine the grain remaining in the sieve as well as the material collected in the pan beneath the sieve. Many storage insects are small and may pass through the sieve. Spread the sieved material and

Table 9.4. Lethal dose (LD, 90% and 99% kill) of phosphine required for two common insects in stored rice at 77°F in 4 days

Insect	Most resistant stage	LD_{90} (ppm)	LD_{99} (ppm)
rice weevil	pupa	32	62
lesser grain borer	egg	19	62

Source: Reed 2006.

Table 9.5. Effect of exposure time at 77°F on lethal dose of phosphine at the most resistant stage of development

Insect	Mortality	Exposure time		
		2 days (ppm)	4 days (ppm)	7 days (ppm)
rice weevil (pupa)	90%	60	32	16
lesser grain borer (egg)	99%	567	70	9

Source: Reed 2006.

use an incandescent lamp for inspection. The heat of the lamp will cause any mobile insects to readily move, making detection easier.

Control of Insects

Active infestations require taking immediate measures to control the pests. Currently five materials are registered in California for controlling storage insects: aluminum phosphide or magnesium phosphide compounds (phosphine), carbon dioxide, sulfuryl fluoride (Profume), diatomaceous earth, and a blend of DE and pyrethrin (e.g., Diatect V).

Phosphine

Phosphine fumigation is usually applied as tablets that release phosphine gas. Grain can be treated with phosphine in the storage without moving the grain. Phosphine is a widely used fumigant for insect control in stored grain. It is applied in solid form; the reaction of metal phosphide (aluminum or magnesium) with water vapor liberates phosphine gas. Other factors being equal, the rate of gas generation increases with temperature. The same volume of phosphine gas is generated in 14 hours at 86°F as in 33 hours at 59°F. Similarly, fumigated cold grain is not ready for shipment as quickly as fumigated warm grain; phosphine takes longer to dissipate from cold rice. At temperatures below 86°F insect control is likely to be reduced. It is ineffectual below 59°F.

Proper timing of phosphine application requires intensive sampling. The objective is to expose the insect to phosphine at a sensitive stage of life, when the population is approaching or exceeded an economic threshold, when the quality of the grain is not compromised, and when environmental conditions maximize efficacy. The most resistant life stages for rice weevil and lesser grain borer are the pupa and egg stages, respectively (table 9.4). Under experimental conditions (Reed 2006), a 99 percent kill at these stages requires 4 days' exposure at 62 parts per million (ppm) for both rice weevil and lesser grain borer. Lethal concentrations are affected by exposure time. To kill 99 percent of lesser grain borer eggs in 2 and 7 days required concentrations of 567 and 9 ppm, respectively (table 9.5). The lethal concentrations to control storage insects generally decreases with increasing temperatures.

The range of concentrations achieved at the label-recommended application rate produces phosphine levels in excess of what is needed to kill rice storage insects. The excessive concentration is intended to compensate for gas leakage from the storage and lack of uniform dispersal within the grain mass. Keep in mind that phosphine circulation within the storage bin depends on air currents within the grain mass. Changes in the air currents over the course of the fumigation can cause phosphine concentrations to vary considerably by location in the storage. Consequently, some parts of the grain mass will be exposed to lower concentrations of phosphine for a shorter time, which may fail to kill all stages of growth of the target insect. Poorly sealed bins and inadequate amounts of phosphine introduced amplify this condition. Although fumigation may reduce the adult population below detection, the pupae may remain viable. A rapid rebound in the insect population after fumigation suggests that lethal concentrations were not maintained long enough or that there was a significant variation in concentration within the storage. In the United States, few insect strains are highly resistant to phosphine. If applied properly phosphine will control all stored rice insects in California.

Phosphine checklist

- A sealed storage bin is needed to keep the phosphine concentration sufficiently high for the required duration to kill all insects. At best, only adult insects are killed in unsealed storages. To hold gas concentrations as long as needed, seal cracks and openings with tape and thick plastic sheets (phosphine migrates through thin plastic).

- Applying phosphine in unsealed storages accelerates the development of pesticide resistance (insects that survive brief or nonlethal applications breed populations that also survive them).

Figure 9.16. Time and carbon dioxide concentration required to kill rice weevil at 65% relative humidity and 70°F. *Source:* Annis and Morton 1997.

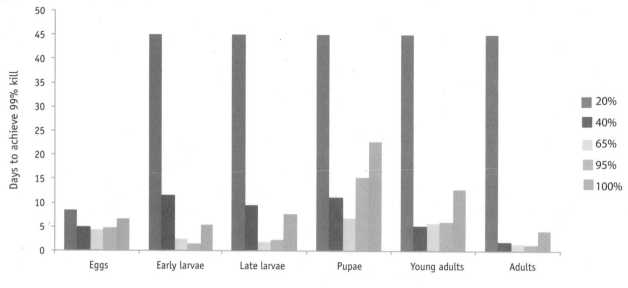

- Temperature within the grain mass affects efficacy of phosphine.

- Insects are most susceptible to phosphine at or above 86°F.

- Minimum exposure time increases with decreasing temperature. Phosphine is ineffective below 59°F.

- Rice must be aerated after the phosphine application for 1 day with fans or 5 days without fans before handling.

- A holding period of 2 days after treatment is required before the rice can be used for human food or stockfeed. Rice can be transported during this period.

- It is illegal and dangerous to fumigate rice during transport.

- Always follow label recommendations.

Carbon Dioxide

Carbon dioxide can be used as a fumigant to control insects in stored rice. It is an expensive control method, and its commercial use is limited to organic operations that store grain of very high value and have few other options.

Carbon dioxide is toxic to insects; but insects will not breathe it in if no oxygen is present in the air (they close their spiracles and go into a state like hibernation). This is why it is recommended that carbon dioxide levels be

Carbon dioxide stored in high-pressure tanks can be used for nonchemical control of insects. *Photo:* Michael Poe.

Figure 9.17. Percent mortality of lesser grain borer exposed for 2 weeks on short (S-102) and medium (M-205) grain rice varieties at 0, 125, 375, and 500 ppm diatomaceous earth at 90°F and 75% relative humidity. *Source:* Fields and Korunic 2000.

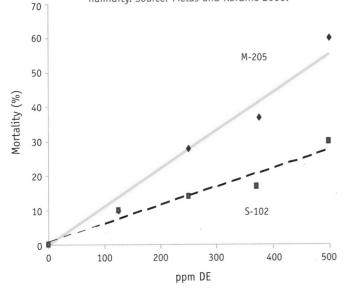

and carbon dioxide to increase susceptibility of pests to phosphine. It has been proven effective against the Angoumois moth and the rice weevil. A cylinderized blend of carbon dioxide and phosphine is marketed under the trade name Ecofume. The handling and delivery of the pressurized gas requires additional training. Safety precautions are similar to those for phosphine fumigations.

Diatomaceous Earth

Diatomaceous earth is an alternative to synthetic insecticides. It is made from the fossilized remains of freshwater or marine unicellular diatoms. Generally, it contains 70 to 90 percent amorphous silicon dioxide, with the remainder being inorganic oxides and salts. It absorbs the cuticular lipids of insects, causing death by desiccation. Relative humidity greater than 80 percent reduces the efficacy of diatomaceous earth because the insect desiccation rate slows. It does not kill grubs inside rice kernels. Disadvantages are that grain is difficult to auger and causes excessive equipment wear.

Diatomaceous earth has been extensively tested for control of stored grain pests (e.g., Fields and Korunic 2000). In direct application studies, exposure of 1 week or less at 300 ppm can produce high levels of mortality in rice weevil. However, satisfactory control of lesser grain borer typically requires higher concentrations and exposure times. The efficacy varies with commercial formulations and commodities treated. Consequently, results on its effectiveness may not be transferable between commodities. There is also evidence that its effectiveness varies between varieties of rice. For example, control of the lesser grain borer was less effective on short grain rice (variety S-102) than on medium grain rice (M-205, fig. 9.17). In M-205, 60 percent mortality of adult lesser grain borers was achieved with 500 ppm, whereas in S-102, the same concentration killed less than 30 percent of the adults. S-102 has trichomes, hairlike structures on the palea and lemma (hull), but M-205 does not; the trichomes may impede the adherence of diatomaceous earth to the rice

increased gradually during application, up to the desired concentration. Higher concentrations of carbon dioxide do not always lead to better control; the optimal concentration depends on the growth stage of the insect (fig. 9.16). Concentrations above 65 percent are lethal to all life stages of grain pests. The recommended concentrations should be maintained for 4 to 5 days at or above 80°F, 7 to 10 days at 60°F, and 14 to 21 days below 60°F. Carbon dioxide is heavier than air and tends to settle to the bottom of the storage, leaving the lowest concentration in the top portions of the bin. To determine the number of days of exposure required, measure the concentrations at the top and bottom of the storage and use the treatment time for the area of lowest concentration. Carbon dioxide is effective only in well-sealed bins.

Carbon Dioxide Plus Phosphine

Treatment with a combination of carbon dioxide and phosphine consists of 50 to 100 ppm phosphine plus 5 percent carbon dioxide at 89°F. Mortality of 100 percent is generally achieved within 24 hours if target concentrations are maintained. This relies on elevated temperature

kernel, or they may slow the lesser grain borers, reducing their exposure. Diatomaceous earth is thought to attach to the cuticle of the lesser grain borer as they disperse downward from the grain surface through the treated layer and into the grain mass. If the beetles do not obtain a lethal dose while passing through the treated layer, diatomaceous earth may not be effective. In practice, higher concentrations may be required in pubescent varieties, and even then control levels may be suboptimal.

Diatomaceous Earth Plus Pyrethrin

Diatomaceous earth plus pyrethrin is certified as an organic insecticide. Pyrethrin is extracted from the seed cases from plants in the chrysanthemum family. It is a neurotoxin that attacks the nervous systems of all insects. The abrasive action of the diatomaceous earth on the insect's exoskeleton facilitates the uptake of pyrethrin.

Sulfuryl Fluoride

Sulfuryl fluoride is considered to be a replacement for methyl bromide in some postharvest fumigation uses. It is a biocide that is lethal to all living organisms with sufficient exposure time and concentration. For this reason, everyone must leave the structure before the fumigation begins and remain outside until the structure is well ventilated for about 24 hours. It is effective on all life stages of postharvest insect pests. Larvae, pupae, and adult insects are highly susceptible, while eggs are more tolerant. Effective dosages for all life stages can be obtained by varying the concentration and exposure time. Lower dosages can be used at higher temperatures because insect metabolic rates increase with temperature. It has been used successfully to control Indian meal moth, confused flour beetle, sawtoothed grain beetle, lesser grain borer, granary weevil, and rice weevil.

Alternative Insect Control Methods

Low Temperature

Lowering the temperature of stored rice can provide control of insect pests. Complete kill of rice weevils in a grain mass under experimental conditions required 16°F for 250 hours or 14°F for 150 hours (Fields and Muir 1996). At a storage temperature of 20°F there was a 30 percent survival rate even after 400 hours (17 days). In general, a temperature below 37°F is required to immobilize insects, and below 23°F is required to kill them (table 9.6).

Expert Systems

An expert system for grain storage is an interactive computer program that incorporates research results with real-time storage bin and weather conditions to assist with management decisions. The system uses a set of biological and physical rules to analyze information supplied by the user and to recommend one or more courses of action to protect the quality of the stored grain. Such systems are vital to an integrated management strategy for storage pests.

Two expert systems for managing stored grain are available in the United States. The Post-Harvest Grain Management (PHGM) was developed in collaboration by Texas A&M, the University of Arkansas, the University of Missouri, and USDA-ARS. This program is a Web-based application that assists the user in developing strategies for bin aeration and pest management. The weather database is updated automatically with data from several weather observation sources, with the option of using historic or real-time information. The system evaluates the on-farm storage capabilities of producers to determine the effectiveness of ambient aeration to control rice grain pests using new and existing aeration and pest data. It is menu driven and allows users to create different scenarios of bin configurations and initial pest infestations to simulate changes in grain temperature and moisture content, along with

Table 9.6. Response of grain storage insects to low temperature

Response	Temperature	Comments
suboptimum	68° to 77°F	slow development
	55° to 68°	development slows or stops
lethal	37° to 55°F	death in days for nonacclimatized insects; movement stops
	14° to 23°F	death in weeks to months if acclimatized
	–13° to–5°F	death in minutes; insects freeze

Source: Fields and Muir 1996.

the resulting pest density and grain damage. The program provides advanced options for analysis of simulation results and economic analysis.

Researchers at Kansas State University, Oklahoma State University, and the USDA-ARS Research Center in Manhattan, Kansas, developed Stored Grain Advisor Pro (SGA Pro), an area-wide IPM program for insects in stored wheat. SGA Pro is decision support software that interprets insect sampling data and provides grain storage managers a report detailing which bins are at risk for insect-caused economic losses. Insect densities can be predicted up to 3 months into the future based on current insect density, grain temperature, and moisture content. The risk-analysis-based program allows reduced sampling frequency and determines fumigation schedules based on projected insect populations rather than on a calendar-based approach. SGA Pro is an improvement over calendar-based management because it ensures the grain is treated only when insects exceed economic thresholds.

Web Sites for Expert Systems

More information on the expert systems and the software is available at the following Web sites:

- Post-Harvest Grain Management (PHGM): http://beaumont.tamu.edu/RiceSSWeb/

- Stored Grain Advisor Pro (SGA Pro): http://www.ars.usda.gov/services/software/download.htm?softwareid=81

References

Abdullah, N., A. Nawawi, and I. Othman. 2000. Fungal spoilage of starch-based foods in relation to its water activity. Journal of Stored Product Research 36(200): 47–54.

Agriculture and Agri-Food Canada, Cereal Research Centre, Winnipeg. Web site http://www.agr.gc.ca.

Annis, P. C., and R. Morton. 1997. The acute mortality effects of carbon dioxide on various life stages of Sitophilus oryzae. Journal of Stored Product Research 33:115.

ARS (U. S. Department of Agriculture Agricultural Research Service). ARS Image Library. ARS Web site, http://www.ars.usda.gov/is/graphics/photos/.

Brooker, D. B., F. W. Bakker-Arkema, and C. W. Hall. 1992. Drying and storage of grains and oilseeds. New York: Van Nostrand Reinhold.

Champagne, E. T, J. F. Thompson, K. L. Bett-Garber, R. G. Mutters, J. A. Miller, and E. Tan. 2004. Impact of storge of freshly harvested paddy rice on milled white rice flavor. Cereal Chemistry 81(4): 444–449.

Cotton, R. T., H. H. Walken, G. D. White, and D. A. Wilken. 1960. Causes of outbreaks of stored grain insects. Kansas Agricultural Experiment Station Bulletin 416.

CSL (Central Services Laboratory, United Kingdom.) Why are mites a problem? CSL Web site, http://www.csl.gov .uk/servicesOverview/foodAnalysis/ miteDetection/whyMitesProblem.cfm.

Degesch America. Pest gallery. Degesch Web site, http://www.degeschamerica.com/pests .html.

Fields, P. G., and Z. Korunic. 2000. The effect of grain moisture content and temperature on the efficacy of diatomaceous earths from different geographical locations against stored product beetles. Journal of Stored Product Research 36:1–3.

Fields, P. G., and W. E. Muir. 1996. Physical control. In B. Subramanyam and D. W. Hagstrum, eds., Integrated management of stored products. Boca Raton, FL: CRC Press.

Flinn, P., D. W. Hagstrum, C. Reed, and T. W. Phillips. 2003. United States Department of Agriculture Agricultural Research Service stored-grain area-wide integrated pest management program. Pest Management Science 59:614–618.

Luh, B. S. 1980. Rice: production and utilization. Westport, CT: AVI.

Mew, T. W., and J. K. Misra, eds. 1994. A manual of rice seed health testing. Los Banos, Philippines: International Rice Research Institute.

Mourier, H., and K. P. Poulsen. 2000. Control of insects and mites in grain using high temperature/short time (HTST) techniques. Journal of Stored Product Research 36:309.

Navarro, S., and R. T. Noyes. 2001. The mechanics and physics of modern grain aeration management. Boca Raton: CRC Press.

Ofuya, T. I., and C. Reichmuth. 2002. Effect of relative humidity on the susceptibility of Callosobruchus maculatus to two modified atmospheres. Journal of Stored Product Research 38:139.

Rajendran, S. 2000. Inhibition of hatching of *Tribolium confusium* by phosphine. Journal of Stored Product Research 36:101.

Reed, C. R. 2006. Managing stored grain to preserve quality and value. St. Paul: American Association of Cereal Chemists International, MN.

Reed, C., and H. Pan. 2000. Loss of phosphine from unsealed bins of wheat at six combinations of grain temperature and grain moisture content. Journal of Stored Product Research 36:263.

Sauer, D. B., ed. 1992. Storage of cereal grains and their products. 4th ed. St. Paul: American Association of Cereal Chemists.

Michael Poe

Quality Changes during Handling and Storage

Quality Factors

- Freshly harvested wet rice may develop off odors and off flavors within even a few hours.

- Flavor and textural properties generally decrease in storage for japonica varieties. This quality loss can be minimized by storing rice at cool temperatures and relatively low moisture content (13 to 14%).

Introduction

The physicochemical properties of rice are influenced, positively and negatively, by storage conditions and duration. The extent to which quality traits are affected is a function of the interaction between variety, storage time, grain temperature, and moisture content. The aging of rice causes changes in head rice yield, water adsorption, grain hardness, volume expansion, pasting properties, texture, and taste. While most of the significant changes occur within the first few months of storage, they can continue indefinitely. Consumers of long grain rice (indica varieties) prefer aged rice and may, in fact, try to

hasten the aging process. In contrast, consumers who prefer japonica rice may go to considerable lengths to slow or halt the physicochemical changes in storage. The degree and rate of change in a number of characteristics may occur under short-term storage conditions and may be affected by drying temperatures. Aging of rice is a complicated process involving physical, chemical, and biological change. The exact mechanisms and biological pathways remain unclear. Proposed mechanisms of aging involve the oxidation of lipids and proteins; the resulting products interact with the carbohydrate fraction and cell membranes to alter the physicochemical properties of the rice. Millers and processors must be aware of the effects of aging on rice quality so they can understand why rice shipped at different times of the year performs differently during processing.

Changes in Taste and Processing Quality

Quality changes over time can improve or diminish the eating and processing characteristics of rice. Changes in rice quality during handling and storage are minimized by cool storage temperatures and are amplified by length of storage time and high water activity. Generalized trends in quality are listed in table 10.1. The degree to which these changes affect rice quality depends on the end user. Alterations in viscosity and gelatinization temperature affect the processing characteristics of rice flour, while changes in textural components, such as stickiness, can affect its value as table rice (fig. 10.1). Aged rice when cooked produces decreased cohesiveness, drier surfaces, larger volumes, and firmer textures; cooking times tend to lengthen as well. Whether these quality changes are desirable depends on the market and the type of rice. Although cooked rice texture is multidimensional, hardness and stickiness govern palatability of rice in north Asian markets. The components of cooked rice flavor begin changing in the early stages of storage. Taste quality can deteriorate shortly after harvest (fig. 10.2). Taste scores for Koshihikari rice dropped 2 points within 7 months of harvest. Palatability was preserved when the rice was held at 56°F. Conversely, aging improves the flavor and textural characteristics of basmati types of rice.

Table 10.1. Changes in selected rice characteristics during storage affecting quality

Characteristic	Effect	Quality Consequences
yellowness	+	Not a problem if rice is stored at cool temperatures; see below.
expanded volume	– or +	Aged rice takes up less water during cooking, which increases the firmness of the cooked rice. Desirable in long grain rice. Undesirable change for premium japonica varieties.
taste	– or +	Effect is market-dependent.
viscosity	–	A measure of the processing quality of rice, particularly rice flour.
chewiness	+	Textural quality not favored by some Asian markets.
stickiness	–	A valuable change particularly in long grain markets; an undesirable change for premium japonica varieties.
B vitamins	–	Long-term storage lessens the nutrient value of brown rice.
reducing sugars	+	Changes the gelatinization properties of rice flour.
enzyme activity	–	Predisposes rice to off flavors and odors.
fatty acids	+	Predisposes rice to off flavors and odors.
amino acids	–	Free amino acids are complexed into proteins. High protein content is related to low taste scores for some varieties.
head rice yield	none	No change or slight increase possible within the first 2 to 3 months of storage.

Key: (+) denotes an increase; (–) denotes a decrease.
Source: Siebenmorgen et al 2004.

Figure 10.1. Changes in texture value of two characteristics of Koshihikari rice. *Source:* Masuro et al. 1997.

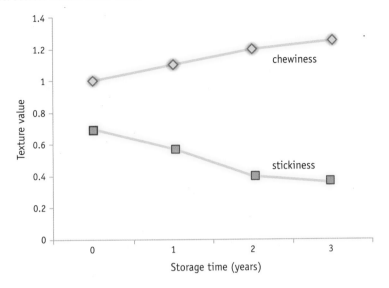

Figure 10.2. Taste scores of Koshihikari rice as related to storage duration and low-temperature storage at 56°F and normal storage at ambient temperature. *Source:* Tani et al. 1964.

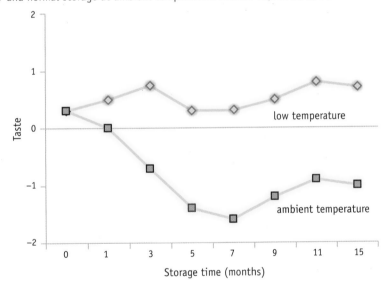

Environmental Factors

Water Activity

The quality of rice once it has been harvested and dried is controlled by three environmental factors: water activity (a_w), temperature, and storage time. Water activity is also referred to as the equilibrium relative humidity (ERH). It represents the ratio of the water vapor pressure in the grain compared to the vapor pressure over pure water. This value multiplied by 100 gives the equilibrium relative humidity. Rice with a water activity of 0.7 has an equilibrium relative humidity of 70 percent. Water activity is a measure of the unbound water in the grain. Water not bound to a constituent of the rice can foster chemical reactions in the rice and support microbial growth.

Water activity is not the same as moisture content. Holding rice at what would appear to be a desirable moisture content may not protect against mold or insect growth. For example,

storing rice at 14 percent moisture content during the cool winter months preserves grain quality. But holding rice at 14 percent moisture content during the warm summer months allows slow growth of mold. Generally, rice storage at a water activity of less than 0.60 ensures negligible microbial activity irrespective of rice temperature (fig. 9.11).

Storage Temperature

Storage temperature impacts the rate of microbial growth, insect growth, and chemical changes in the rice grain. At temperatures around 85°F detrimental changes occur in a few weeks. Commercially, rice should be stored below 60° F to achieve low levels of quality loss. Under experimental conditions, favorable rice quality has been preserved for years at subzero temperatures. In the United States storage temperatures generally range from 50° to 95°F, rice moisture content varies from 10 to 15 percent, and the typical storage duration is between 2 and 24 months.

Storage Duration

Time is critical in controlling the amount of quality loss. Rice with a high water activity may be held for short periods of time. For example, freshly harvested high-moisture rice can be held for several hours without developing off odors, but after several days of nonaerated storage, taste quality will be permanently damaged.

Quality Attributes

Even with suitable handling, drying, and storage conditions, rice quality changes during processing and storage. Some quality changes during storage, such as in free fatty acid production and water absorption ratio, are thought to be caused by enzymatic reactions involving protein, starch, and lipid. Generally the outer layers of the rice kernel are more susceptible to these reactions than the endosperm (fig. 10.3). The bran layer (pericarp, tegmen, and aleurone) is intact in brown rice. The enzymatically active bran layer accounts for the tendency of brown

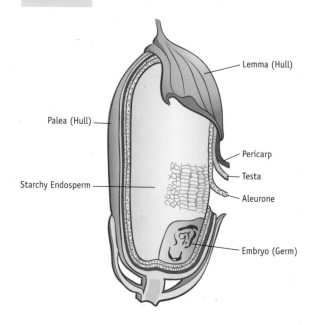

Figure 10.3. Cross-section of the rice kernel.

rice to become rancid rather quickly if stored improperly.

Yellowing

Rice yellows during storage, even without mold development, apparently because of a nonenzymatic browning process known as the Maillard reaction. The rate of yellowing is slow at typical safe storage conditions (fig. 10.4). For example, at a water activity of 0.6 and a temperature of 75°F, corresponding to a moisture of about 12 percent, yellowing is just detectable after 200 days of storage. But as the temperature and water activity of the rice increase, yellowing can become a serious problem. At water activity of 0.6 and a temperature of 85°F yellowing is noticeable after 90 days of storage; if water activity is increased to 0.8 yellowing is detectable after only 45 days. Mold growth can also cause yellowing, also called stack burn, because the fungi produce compounds that stain the rice. Mold can grow when rice has high water activity (increased rice moisture, such as that caused by a roof leak following a rain). This growth results in increased temperature and moisture content, which promote yellowing caused by the Maillard reaction.

Protein

Research has demonstrated that protein content affects texture, tenderness, and cohesiveness of cooked rice (Hamaker 1994). Proteins are most concentrated in the outer layers of the rice kernel (see fig. 10.3 and table 10.2), but significant amounts are also present in the endosperm. Total protein does not change during storage; however, the chemical properties of the proteins can be altered. Free amino acids, for example, in stored rice decrease with time. The decline is associated with textural changes such as reduced adhesiveness in the cooked rice.

One mechanism by which protein influences texture is hypothesized to involve the regulation of water diffusion into the starch granule, which consequently alters the water and time requirements for cooking. Also, protein may impede starch gelatinization. This is particularly important in the use of rice flour as a thickening agent in such products as baby food (see chapter 1, "Rice Quality in the Global Market"). During storage the interaction of protein with starch may also alter the gelatinization temperature of the starch and the texture of cooked rice. The brewing industry is reluctant to use rice stored for long periods because it requires higher temperatures for gelatinization and converts slowly to fermentable sugars. Rice textural properties change significantly in the months following harvest. Many of the functional changes that occur during storage that influence cooking properties, such as cooking time, water uptake, and stickiness, are thought to be caused by protein-starch interactions.

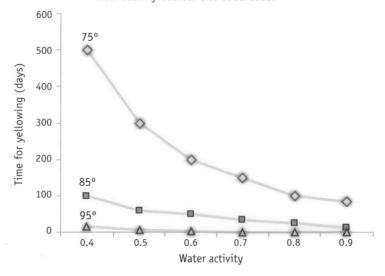

Figure 10.4. Onset of discernable yellowing in Australian-grown Calrose rice caused by storage temperature and water activity. *Source:* Gras et al. 1989.

Table 10.2. Approximate chemical composition of the outer layer, endosperm, and entire kernel of rice

Constituent	Unit	Outer layer	Endosperm	Entire kernel
starch	%	60.68	92.00	90.68
total sugars	%	2.92	0.18	0.38
fiber	%	1.47	0.22	0.28
protein N	g N/100 g rice	2.49	1.25	0.90
total lipids	%	4.44	0.45	0.66
free fatty acids	%	1.34	0.15	0.21
thiamine	%	0.80	0.05	0.08
niacin	mg/100 g	0.08	0.02	0.02
riboflavin	mg/100 g	9.27	0.89	1.26
pyridoxine	mg/100 g	1.19	0.08	0.13
calcium	%	0.36	0.01	0.02
iron	%	0.03	—	0.00
phosphorus	%	1.02	0.10	0.14

Source: Chrastil 1994.

Carbohydrates

The rice kernel is approximately 90 percent carbohydrate (e.g., starch or sugars; table 10.2). Grain respiration is very slow in storage, so very little of the carbohydrate content is lost, although loss can become significant at moisture contents greater than 14 percent and at high temperatures. Similarly, total starch content does not change appreciably during storage, although small changes in starch properties have been observed. Reducing sugars (e.g., maltose) increase and nonreducing sugars (e.g., sucrose) decrease during storage when subjected to temperatures of 77°F or greater (fig. 10.5). Enzymes such as amylase initiate starch synthesis and degradation processes in the rice grain during storage. For example, a decrease in amylase activity during storage causes an increase in the gelatinization temperature. Major changes in cooked rice hardness, gel consistency, and viscosity values require approximately 3 months to occur. Therefore, textural evaluation should be done on samples aged at least 3 months. High-amylose varieties have the greatest textural changes in storage. Cold storage reportedly preserves cohesiveness, a desired trait in some Asian markets.

Lipids

Most of the lipids in rice grains are concentrated in the germ and bran layer. The endosperm contains only a fraction of the total lipid content. Lipolytic enzymes, specifically lipases, catalyze the hydrolysis of kernel lipids, causing the formation of free fatty acids. These enzymes exist naturally in rice and may also come from microbial growth. Lipases are compartmentalized in the testa and aleurone layers, and lipids are in the germ. Therefore, these reactions occur principally in the outer layer of the rice kernel, where lipids are concentrated. Dehulling rice disrupts the outer layers, causing lipids to diffuse and make contact with the lipases. Microbially produced lipases located on the kernel surface also can come in contact with kernel lipids (fig. 10.6).

Research results suggest that lipid degradation is responsible for much of the deterioration of rice in storage. The appearance of free fatty acids, which accelerates at high temperature, is an indicator of lipid degradation (fig. 10.7). Living rice grains are constantly exposed to toxic by-products of metabolism (e.g., superoxides, hydrogen peroxide, and hydroxyl radicals). Lipid oxidation induced by these compounds compromises cell membrane

Figure 10.5. Content of reducing sugars as related to storage duration at different temperatures. Low-temperature storage was at 55°F and 70% relative humidity. *Source:* Tani et al. 1964.

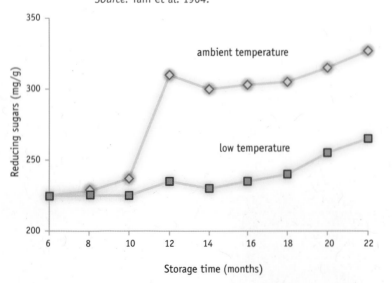

Figure 10.6. Routes of hydrolytic and oxidative deterioration in rice bran lipids. *Source:* Champagne 1994.

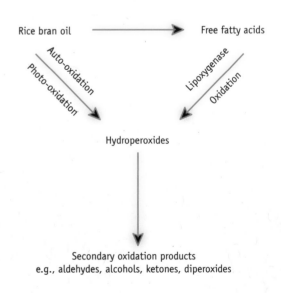

Figure 10.7. Changes in free fatty acids during storage at two temperatures. Low temperature equals 41°F, and high equals 95°F. *Source:* Zhou et al. 2002.

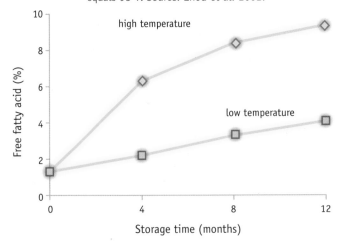

peroxides formed during this reaction yield subsequent products that produce undesirable odors and flavors.

Odor and Flavor

The odor and flavor of rice when cooked changes with time even when stored in airtight conditions. Aldehydes and ketones are the dominant source of the off flavors, although other undesirable compounds are also detected (e.g., sulfur dioxide, hydrogen sulfide). Most of the detected volatile compounds are products of lipid, amino acid, or vitamin decomposition. Well-polished rice (12 percent moisture content) retains its flavor longer than less-polished rice because the outer layers of the kernel contain the highest levels of oxidizable compounds that contribute to off flavors.

Milling Yield in Storage

Studies have shown that milling yield does not change during storage. Warehouse operators often observe an increase based on samples analyzed with FGIS procedures. This effect may be due to a decrease in sample moisture over the storage period (milling yields increase slightly as sample

integrity. For example, catalase, an antioxidant enzyme, decomposes hydrogen peroxide to water and oxygen, rendering oxidizing compounds harmless. Catalase activity in stored rice declines steadily over time (fig. 10.8). Low-temperature storage reduces the loss of activity compared to storage at high temperatures. The freshness of rice can be determined using a color reaction technique to measure catalase activity.

The rate of free fatty acid formation in brown rice increases with greater surface disruption (e.g., hulling), increased rice moisture content, greater amounts of microbial infection, and higher grain temperature. Approximately 30 percent of the kernel oil content can be converted to free fatty acids within 1 week under high humidity and temperature (fig. 10.9). An enzyme called lipoxygenase, found in the germ, catalyzes the oxidation of fatty acids to peroxides that in turn produce other compounds (e.g., aldehydes, ketones, and carboxylic acids). These products of lipid breakdown result in the development of off odors and off flavors associated with "spoiled" rice.

Rice lipids are also able to undergo non-enzymatic oxidation. Endogenous metal ions or those introduced in the shelling process, as well as light energy and heat, catalyze non-enzymatic oxidation. Similar to enzymatic oxidation,

Figure 10.8. Changes in catalase activity in rice during storage. Low-temperature storage was at 55°F and 70% relative humidity. *Source:* Tani et al. 1964.

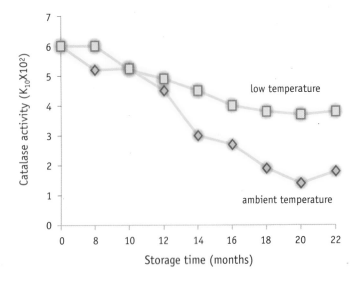

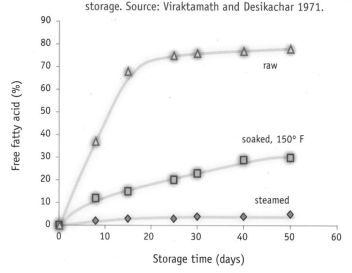

Figure 10.9. Free fatty acid formation in bran from raw ($\triangle$), soaked (150°F, 3 hr, $\square$), and steamed (212°F, 5 min, $\diamondsuit$) during storage. Source: Viraktamath and Desikachar 1971.

moisture decreases). For example, M-202 rice milled at 12 percent moisture had 2 percentage points increase in head rice yield compared with the same batch milled at 13.5 percent moisture (table 10.3). Long grain rice shows a greater effect of moisture on milling yield, and short grain rice shows a smaller effect compared with medium grain rice.

Predicting Storability

Another means of evaluating the integrity of stored grain is based on dry matter loss. As a rule of thumb, stored grain can have 0.5 percent dry matter loss before deterioration affects market grade. The 0.5 percent guideline and its dependence on moisture and temperature have been used in recommendations to grain storage managers by the USDA. Dry matter losses in storage due to grain respiration are difficult to quantify. However, the storage time required to lose 0.5 percent dry matter in rice can be estimated using measured respiration rates (CO_2 evolution). In the absence of mold and insect infestation, paddy rice at 68°F at 15 percent moisture content would lose 0.5 percent of its dry matter weight in 41 months (table 10.4). At the same temperature, rice at 18 percent moisture content requires 2.5 months for the same weight loss. A 1 percent increase in grain moisture more than doubles the respiration rate and reduces the time for discernible weight loss by more than one-half at a given temperature. In well-managed rice stored at less than 14 percent moisture content the respiration loss is negligible during

Table 10.3. Effect of sample moisture on milling yield for California-grown M-202 rice

Sample moisture (% wb)	Head rice yield (%)	Total rice yield (%)
13.5	50.2	65.8
12.0	52.0	67.2
Milling quality increase	1.8	1.4

Source: Thompson et al. 1990.

Table 10.4. Projected storage time for rice based on 0.5% dry matter loss

Temperature (F°)	Months in storage			
	Moisture content (%)			
	15	16	17	18
50			41	11
68	41	13	6	2.5
77	20	5	2	0.8
86	8	1.7	0.7	0.1

Source: Adapted from Hosokawa 1995 and Teter 1981.

the typical duration of storage in California. Nonetheless, isolated wet spots in the bins results in a rapid increase in rice respiration with an associated temperature rise. This creates an ideal environment for mold growth.

Rapid Chemical Evaluation of Grain Freshness

Since quality change is an unavoidable consequence of storage, the Japan Millers Association has established rapid testing procedures for evaluating the quality of stored rice. Two main procedures used are as follows.

1. Enzyme activity. Assay to determine peroxidase activity:

    - Place 2 g of rice in a 1 percent guaiacol solution and shake.

    - Add a few drops of 3 percent hydrogen peroxide solution and shake.

    - Evaluate by the color of the grain. A dark color implies high peroxidase activity that indicates fresh rice.

2. Increased fat acidity. Perform a colorimetric assay to determine fatty acid content based on the pH of a water and rice solution:

    - Make reagent: 0.1 g methyl red and 0.3 g bromothymol in 150 ml of ethanol.

    - Dilute reagent to a 2 percent solution with distilled water.

    - Place 5 g of rice in 10 ml of dilute reagent (acidity indicator).

    - Observe color change: fresh = green, old = yellow or orange.

References

Champagne, E. T. 1994. Brown rice stabilization. In W. E. Marshall and J. I. Wadsworth, eds., Rice science and technology. Boca Raton, FL: CRC Press.

Chrastil, J. 1994. Effect of storage on the physicochemical properties and quality factors of rice. In W. E. Marshall and J. I. Wadsworth, eds., Rice science and technology. Boca Raton, FL: CRC Press.

Daniels, M. J., B. P. Marks, T. J. Siebenmorgen, R. W. McNew, and J. F. Meullenet. 1998. Effects of long grain rough-rice storage history on end-use quality. Journal of Food Science 63(5): 832.

Gras, P. W., H. J. Banks, M. L. Bason, and L. P. Arriola. 1989. A quantitative study of the influences of temperature, water activity, and storage atmosphere on the yellowing of milled rice. Journal of Cereal Science 9:77–89.

Hamaker, B. R. 1994. The influence of rice protein on rice quality. In W. E. Marshall and J. I. Wadsworth, eds., Rice science and technology. New York: Marcel Dekker.

Hosokawa, A., ed. 1995. Rice post-harvest technology. Tokyo, Japan: The Food Agency, Ministry of Agriculture, Forestry, and Fisheries.

Masuro, N., N. Manabu, and I. Shoei. 1997. Relation between rice quality and palatability and storage conditions of Koshihikari. Bulletin of the Toyama Agricultural Research Center 17:51–64.

Matsukura, U., K. Shigenoh, and M. Michiko. 2000. Method for measuring the freshness of individual rice grains by means of a color reaction of catalase activity. Journal of the Japanese Society for Food Science and Technology 47(7): 523–528.

Minato, K., M. Gono, H. Yamaguchi, Y. Kato, and T. Osawa. 2005. Accumulation of N-(hexanoyl) lysine, an oxidative stress biomarker in rice seeds during storage. Bioscience, Biotechnology & Biochemistry 69(9): 1806–1810.

Mutters, R. G., J. F. Thompson, and R. E. Plant. 2007. Crop management and environmental effects on rice milling quality and yield. Report to the California Rice Research Board.

Sauer, D. B., ed. 1992. Storage of cereal grains and their products. 4th ed. St. Paul: American Association of Cereal Chemists.

Seib, P. A., P. Fost, A. Sukabdi, V. G. Rao, and R. Burroughs. 1980. Spoilage of rough rice measured by evolution of carbon dioxide. Proceedings of the Grain Post-Harvest Workshop, Kuala Lumpur, Malaysia. 75–94.

Siebenmorgen, T. J., and J. F. Meullenet. 2004. Impact of drying, storage, and milling on rice quality and functionality. In E. T. Champagne, ed. Rice chemistry and technology. St. Paul, MN: American Association of Cereal Chemists.

Tani, T. S., S. Chikubu, and T. Iwasaki. 1964. Changes in chemical properties of brown rice during the low temperature storage. Japanese Society of Food and Nutrition 16:436.

Teo, C. H., A. A. Karim, P. B. Cheah, M. H. Norziah, and C. C. Seow. 2000. On the roles of protein and starch in the aging of non-waxy rice flour. Journal of Food Chemistry 69(3): 229–236.

Teter, N. C. 1981. Grain storage. Laguna, Philippines: SEARCA College.

Thompson, J. F., J. Knutson, and B. Jenkins. 1990. Analysis of variability in rice milling appraisals. Applied Engineering in Agriculture 6(2): 194–198.

Tsugita, T., O. Takeo, K. Hiromichi, and H. Kato. 1983. Cooking flavor and texture of rice stored under different conditions. Agricultural and Biological Chemistry 47(3): 543–549.

Viraktamath, C. S., and H. S. R. Desikachar. 1971. Inactivation of Lipase in rice bran in Indian rice mills. Journal of Food Science and Technology 8:70–74.

Zhou, Z., K. Robards, S. Helliwell, and C. Blanchard. 2002. Ageing of stored rice: Changes in chemical and physical attributes. Journal of Cereal Science 35:65–78.

English–Metric Conversions

English	Conversion factor for English to metric	Conversion factor for metric to English	Metric
Length			
inch (in)	2.54	0.394	centimeter (cm)
foot (ft)	0.3048	3.28	meter (m)
foot per minute (fpm)	0.0058	196.85	meter per second
Area			
acre (ac)	0.4047	2.47	hectare (ha)
square foot (ft²)	0.0929	10.764	square meter (m²)
Volume			
bushel (bu)	0.035	28.37	cubic meter (m³)
gallon (gal)	3.785	0.26	liter (l)
cubic foot (ft³)	28.317	0.353	liter (l)
cubic foot per minute (cfm)	0.000472	2118.88	cubic meter per second ($m^3 \cdot s^{-1}$)
cubic foot per minute per hundredweight (cfm/cwt)	0.0006243	1601.85	cubic meter per kilogram
cubic foot per minute per pound (cfm/lb)	0.0010405	961.11	cubic meter per second per kilogram ($m^3 \cdot s^{-1} \cdot kg^{-1}$)
cubic yard (yd³)	0.765	1.307	cubic meter (m³)
Mass			
ounce (oz)	28.35	0.035	gram (g)
pound (lb)	0.454	2.205	kilogram (kg)
ton (T)	0.907	1.1	metric ton (t)
pound per bushel (lb/bu)	12.872	0.0777	kilogram per cubic meter
pound per acre (lb/ac)	1.12	0.89	kilogram per hectare (kg/ha)
hundredweight per acre (cwt/ac)	0.112	8.93	metric ton per hectare
Miscellaneous			
Fahrenheit (°F)	°C = (°F – 32) ÷ 1.8	°F = (°C x 1.8) + 32	Celsius (°C)
pound per square inch (psi)	6.89	0.145	kilopascal (kPa)
horsepower (hp)	0.746	1.34	kilowatt (KW)

Index